ANALOG
DIGITAL AND
MICROPROCESSOR
ELECTRONICS

The Author

Er. Nishit Mathur obtained B.Tech (Electronics) from Rama Rao Adik Institute of Technology, Mumbai University, Mumbai. He has published two dozen books. Er. Mathur has more than twelve years experience in e- learning, content designing, project management, digital content writing and editing for School and Universities

ANALOG DIGITAL AND MICROPROCESSOR ELECTRONICS

Author

Er. Nishit Mathur

Kruger Brentt

P u b l i s h e r s

2 0 2 5

Kruger Brentt Publishers UK. LTD.
Company Number 9728962

Regd. Office: 68 St Margarets Road, Edgware, Middlesex HA8 9UU

Library of Congress Cataloging-in-Publication Data

ISBN 978-1-78715-023-2 (Hardbound)

For information on all our publications visit our website at http://krugerbrentt.com/

Preface

Digital electronics study how networks of semiconductor devices such as transistors perform signal-processing tasks. Examples of such tasks include generating and amplifying speech or music, TV broadcasting and displaying, cell phone and satellite communications. Students learn how to design sophisticated electronic microchips to perform these tasks in a variety of electronic systems. The digital nature of electronic signals offers a convenient, compact and noise-free representation of information. Digital signals can be easily stored in an electronic memory and can be easily understood by digital microprocessors. Examples of engineering problems in digital electronics are: how to efficiently perform arithmetic operations with digital signals on a microprocessor, how to communicate data without losing information, and how to design a reusable reconfigurable digital processor. The analog nature of electronic signals is of importance as the real world is analog, and because in modern microchips even digital circuits exhibit analog behaviour. Examples of engineering problems in analog electronics are: how to efficiently represent an analog signal such as an image recorded by a digital camera in a digital format so that it can be stored in a digital memory or processed by a microprocessor; how to send large amounts of information such as high-definition video data from one microchip to another quickly; how to send data such as a text message to a cell phone wirelessly in the presence of interference; and how to design a pacemaker or neural implant to function inside a human body.

Digital electronics, digital technology or digital circuits are electronics that operate on digital signals. In contrast, analog circuits manipulate analog signals whose performance is more subject to manufacturing tolerance, signal attenuation and noise. Digital techniques are helpful because it is a lot easier to get an electronic device to switch into one of a number of known states than to accurately reproduce a continuous range of values. Digital electronic circuits are usually made from large assemblies of logic gates (often printed on integrated circuits), simple electronic representations of Boolean logic functions.

The present book entitled Analog Digital and Microprocessor Electronics contains eleven chapters covering all related disciplines. These chapters include oscillators, power amplifier, BJT and field effect transistor, data and number representation, memory circuits, programmable logic devices, combinational logic circuit, synchronous sequential logic and microprocessors. This book is an asset for digital electronics engineer, digital circuit design engineer, digital integrated circuit design engineer, B.E., B.Tech, M.Sc. (Computer Science/ IT), M.Sc. (Physics), Electronics, BCA and MCA students.

Constructive suggestions of readers especially the teaching faculty and students are invited for further improvement of the book. Their esteemed suggestions will be duly acknowledged in the future editions. I am highly thankful to Publishers Kruger Brentt Publisher, London (UK) for his interest shown in publishing of the book so efficiently and promptly.

Er. Nishit Mathur

Contents

Configuration – Common-Emitter (CE) Configuration – Common Collector (CC) Configuration – Comparison Of Three Configurations – Operating Point (Quiescent, Q Or Silent Point) – Q-Point Value – Different Operating Conditions of Transistor – Introduction – Diode Resistor Transistor Model – Current Source Model – Transistor Biasing – Introduction – Design – Mathematical Approach – Need for Biasing of Transistor – Bad Circuits – Faithful Amplification – Base Resistor/Fixed Bias Circuit – Biasing With Feedback Resistor – Common-Emitter Amplifier without Feedback – Self Biasing – Voltage or Potential Divider Biasing – Battery Bias Stabilisation – Load Current – Bias Compensation – Design of Biasing Circuit – Circuit – FET – Introduction – Construction of JEFT – Biasing of FET – Working Principle of N- FET – Current Control – Static Characteristic of FET – Output or Drain Characteristic – (a) Drain Characteristic with Shorted-Gate – Transfer Characteristic of JFET – Merits/Demerits – The Main Drawback of JFET – FET parameters – FET applications – Low Noise Amplifier – Buffer Amplifier – Cascode Amplifier – Analog Switch – Chopper – Multiplexer – Current Limiter – Phase Shift Oscillators – Small Signal Bjt Amplifier – Introduction – Characteristic – Analysis of Transistor Amplifier.

Oscillators

Introduction

An oscillator must have the following three elements

☆ Oscillatory circuit or element.

☆ Amplifier.

☆ Feedback network.

The oscillatory circuit or element, also called the tank circuit, consists of an inductive coil of inductance L connected in parallel with a capacitor of capacitance C. The frequency of oscillation in the circuit depends upon the values of L and C. The actual frequency of oscillation is the resonant or natural frequency and is given by the expression

f = 1 / 2∏√LC Hz,

where, L is inductance of coil in Henrys, and C is the capacitance of capacitor in farads.

The electronic amplifier receives dc power from the battery or dc power supply and converts it into A.C. power for supply to the tank circuit. The oscillations occurring in the tank circuit are applied to the input of the electronic amplifier. Because of the amplifying properties of the amplifier, we get increased output of these oscillations. This amplified output of oscillations is because of dc power supplied from the external source (a battery or power supply). The output of the amplifier can be supplied to the tank circuit to meet the losses.

The feedback network supplies a part of output power to the tank or oscillatory circuit in correct phase to aid the oscillations. In other words, the feedback circuit provides positive feedback. The block diagram of oscillator is depicted in Figure 1.1.

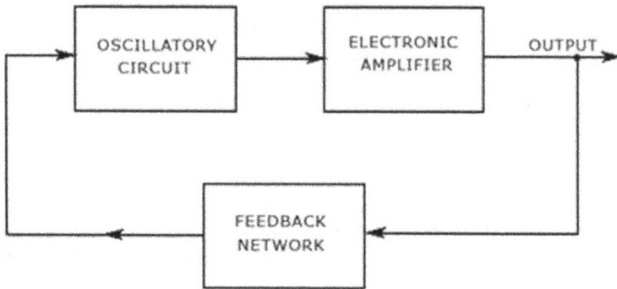

Figure 1.1: Block Diagram of an Oscillator.

Types of Transistor Oscillators

Introduction

A transistor can be operated as an oscillator for producing continuous undamped oscillations of any desired frequency if tank (or oscillatory) and feedback circuits are *property* connected to it. All oscillators under different names have similar function *i.e.* they generate continuous undamped output. However, they differ in methods of supplying energy to the tank or oscillatory circuit to meet the losses and the frequency ranges over which they are used. (The frequency spectrum over which oscillators are employed to produce sinusoidal signals is extremely wide (from less than 1 Hz to many GHz)) However, no single oscillator design is practical for generating signals over this entire range. Instead, a variety of designs are employed, each of which generates sinusoidal outputs most advantageously over various portions of the frequency spectrum. Oscillators, which use inductance-capacitance (L-C) circuits as their tank or oscillatory circuits, are very popular for generating high-frequency (*e.g.* 10 kHz to 100 MHz) outputs.

The most widely used LC oscillators are the Hartley and Colpitt's oscillators. Although they slightly differ from one another in their electronic *circuitry* they have virtually identical frequency ranges and frequency stability characteristics. However, such oscillators are not suitable for generating low-frequency sinusoidal outputs. This is due to the fact that some components needed in the construction of low-frequency LC resonant circuits are too bulky and heavy. So resistor-capacitor (R-C) oscillators are generally employed for generating low-frequency (from 1 Hz to about 1 MHz) sinusoidal signals.

Two most common R-C oscillators are the Wien bridge arid phase-shift types. Other less frequently used oscillators are the crystal oscillators and the negative resistance oscillators. The operating frequency ranges of various types of most commonly used oscillators are given below:

Wien bridge oscillator	1 Hz	—	1 MHz
Phase shift oscillator	1 Hz	—	10 MHz
Hartley oscillator	10 kHz	—	100 MHz
Colpitt's oscillator	10 kHz	—	100 MHz
Negative resistance oscillator		>	100 MHz
Crystal oscillator		Fixed frequency	

Phase Shift Oscillators

The circuit is drawn to show clearly the amplifier and feedback network (Figure 1.2). The circuit consists of a common source FET amplifier followed by a three-section R-C phase shift network. The amplifier stage is self-biased with a capacitor bypassed source resistor R_s and a drain bias resistance R_D. The output of the last section is supplied back to the gate. If the loading of the phase-shift network on the amplifier can be assumed to be negligible, a phase shift of 180° between the amplified output voltage V_{out} and the input voltage V_{in} at the gate is produced by the amplifier itself. The three-section R-C phase shift network produces an additional phase shift, which is a function of frequency and equals 180° at some frequency of operation. At this frequency, the total phase shift from the gate around the circuit and back to the gate will be exactly zero. This particular frequency will be the one at which the circuit will oscillate provided that the magnitude of the amplification is sufficiently large. In a FET phase-shift oscillator voltage series feedback that is, feedback voltage proportional to the output voltage V_{out} and supplied in series with the input signal at the gate is used.

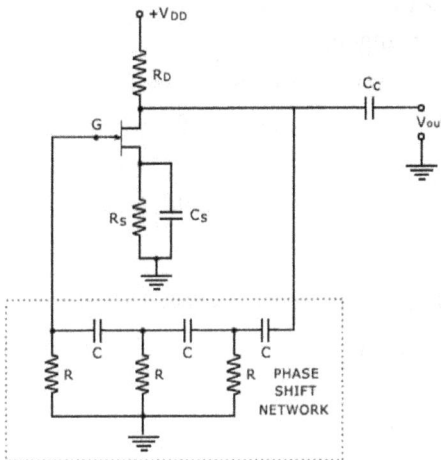

Figure 1.2: Basic Circuit of a FET Shift Oscilattor.

The frequency of the oscillator output depends upon the values of capacitors C and resistors R used in the phase shift network. Using basic RC circuit analysis technique, it can be shown that the network phase shift is 180° when

$$X_c = \sqrt{6}\,R$$

$$1\,/\,2\Pi fc = \sqrt{6}\,R$$

$$f = 1\,//\,2\Pi\,R\,c\,\sqrt{6}$$

The frequency can be adjusted over a wide range if variable capacitors are used. As well as phase shifting, the R-C network attenuates the amplifier output. Network analysis shows that when the necessary phase shift of 180° is obtained, this network attenuates the output voltage by a factor of 1/29. This means that the amplifier must have a voltage gain of 29 or more. When the amplifier voltage gain is 29 and feedback factor of R-C network, $\beta = 1/29$ then the loop gain is $\beta A = 1$, the amplifier phase shift of −180° combined with the network phase shift of +180° gives a loop phase shift of zero. Both of these conditions are necessary to satisfy the Barkhausen criteria. If the amplifier gain is much greater than 29, the oscillator output waveform is likely to be distorted. When the gain is slightly greater than 29, the output is usually a reasonably pure sinusoidal.

The advantages and disadvantages of phase shift oscillators are given below:

Advantages

- ☆ It is a cheap and simple circuit as it contains resistors and capacitors (not bulky and expensive high-value inductors).
- ☆ It provides good frequency stability.
- ☆ The phase shift oscillator circuit is much simpler than the Wien bridge oscillator circuit because it does not need negative feedback and the stabilization arrangements.
- ☆ The output is sinusoidal that is quite distortion free.
- ☆ They have a wide frequency range (from a few Hz to several hundred kHz).
- ☆ They are particularly suitable for low frequencies; say of the order of 1 Hz, as these frequencies can be easily obtained by using R and C of large values.

Disadvantages

- ☆ The output is small. It is due to smaller feedback.
- ☆ It is difficult for the circuit to start oscillations as the feedback is usually small.
- ☆ The frequency stability is not as good as that of Wien bridge oscillator.
- ☆ It needs high voltage (12 V) battery so as to develop sufficiently large feedback voltage.

Applications

FET phase-shift oscillator is used for generating signals over a wide frequency range. The frequency may be varied from a few Hz to 200 Hz by employing one set of resistors with three capacitors ganged together to vary over a capacitance range in the 1 : 10 ratio. Similarly, the frequency ranges of 200 Hz to 2 kHz, 2 kHz to 20 kHz and 20 kHz to 200 kHz can be obtained by using other sets of resistors.

RC Phase Shift Oscillator

In amplifiers, we saw that a single stage amplifier will produce 180° of phase shift between its output and input signals when connected in a class-A type configuration. For an oscillator to sustain oscillations indefinitely, sufficient feedback of the correct phase, ie "Positive Feedback" must be provided with the amplifier being used as one inverting stage to achieve this. In an RC Oscillator, the input is shifted 180° through the amplifier stage and 180° again through a second inverting stage giving us "180° + 180° = 360°" of phase shift which is the same as 0° thereby giving us the required positive feedback. In other words, the phase shift of the feedback loop should be "0".

In a Resistance-Capacitance Oscillator or simply an RC Oscillator, we make use of the fact that a phase shift occurs between the input to an RC network and the output from the same network by using RC elements in the feedback branch, for example.

RC Phase-Shift Network

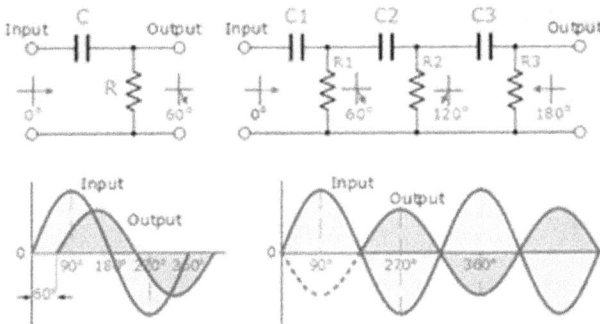

The circuit on the left shows a single resistor-capacitor network and whose output voltage "leads" the input voltage by some angle less than 90°. An ideal RC circuit would produce a phase shift of exactly 90°. The amount of actual phase shift in the circuit depends upon the values of the resistor and the capacitor, and the chosen frequency of oscillations with the phase angle (Φ) being given as:

$$X_C = \frac{1}{2\pi f C} \qquad R = R,$$

$$Z = \sqrt{R^2 + (X_C)^2}$$

$$\therefore \phi = \tan^{-1} \frac{X_C}{R}$$

In our simple example above, the values of R and C have been chosen so that at the required frequency the output voltage leads the input voltage by an angle of about 60°. Then the phase angle between each successive RC section increases by another 60° giving a phase difference between the input and output of 180° (3 × 60°) as shown by the following vector diagram.

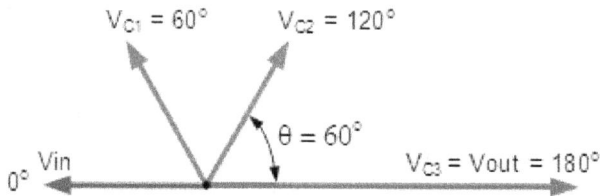

Then by connecting together three such RC networks in series, we can produce a total phase shift in the circuit of 180° at the chosen frequency and this form the bases of a "phase shift oscillator" otherwise known as a **RC Oscillator** circuit.

We know that in an amplifier circuit either using a Bipolar Transistor or an Operational Amplifier, it will produce a phase-shift of 180° between its input and output. If an RC phase-shift network is connected between this input and output of the amplifier, the total phase shift necessary for regenerative feedback will become 360°, ie. the feedback is "in-phase". Then to achieve the required phase shift in an RC oscillator circuit is to use multiple RC phase-shifting networks.

RC Oscillator Circuit

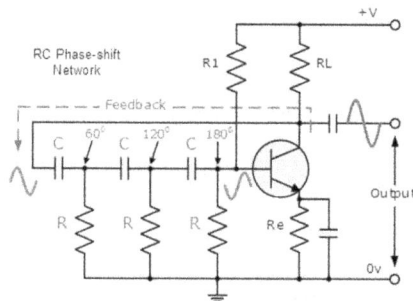

Figure 1.3: RC Oscillator Circuit.

The **RC Oscillator** which is also called a **Phase Shift Oscillator** produces a sine wave output signal using regenerative feedback from the resistor-capacitor combination (Figure 1.3). This regenerative feedback from the RC network is due to the ability of the capacitor to store an electric charge, (similar to the LC tank circuit). This resistor-capacitor feedback network can be connected as shown above to produce a leading phase shift (phase advance network) or interchanged to produce a lagging phase shift (phase retard network) the outcome is still the same as the sine wave oscillations only occur at the frequency at which the overall phase-shift is 360°. By varying one or more of the resistors or capacitors in the phase-shift network, the frequency can be varied and generally this is done using a 3-ganged variable capacitor. If all the resistors, R and the capacitors, C in the phase shift network are equal in value, then the frequency of oscillations produced by the RC oscillator is given as:

$$f_r = \frac{1}{2\pi RC\sqrt{2N}}$$

where,

f is the Output Frequency in Hertz

R is the Resistance in Ohms

C is the Capacitance in Farads

N is the number of RC stages. (in our example N = 3)

Since the resistor-capacitor combination in the **RC Oscillator** circuit also acts as an attenuator producing an attenuation of -1/29th (Vo/Vi = β) per stage, the gain of the amplifier must be sufficient to overcome the losses and in our three mesh network above the amplifier gain must be greater than 29. The loading effect of the amplifier on the feedback network has an effect on the frequency of oscillations and can cause the oscillator frequency to be up to 25% higher than calculated. Then the feedback network should be driven from a high impedance output source and fed into a low impedance load such as a common emitter transistor amplifier but better still is to use an **Operational Amplifier** as it satisfies these conditions perfectly.

Op-amp RC Oscillator

When used as RC oscillators, Operational Amplifier RC Oscillators are more common than their bipolar transistors counterparts. The oscillator circuit consists of a negative-gain operational amplifier and a three section RC network that produces the 180° phase shift. The phase shift network is connected from the op-amp's output back to its "non-inverting" input as shown in Figure 1.4.

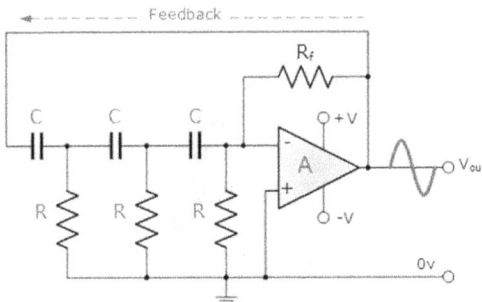

Figure 1.4: Operational Amplifier RC Oscillators.

As the feedback is connected to the non-inverting input, the operational amplifier is therefore connected in its "inverting amplifier" configuration which produces the required 180° phase shift while the RC network produces the other 180° phase shift at the required frequency (180° + 180°). Although it is possible to cascade together only two RC stages to provide the required 180° of phase shift (90° + 90°), the stability of the oscillator at low frequencies is poor.

One of the most important features of an **RC Oscillator** is its frequency stability which is its ability to provide a constant frequency output under varying load conditions. By cascading three or even four RC stages together (4 × 45°), the stability of the oscillator can be greatly improved. *RC Oscillators* with four stages are generally used because commonly available operational amplifiers come in quad IC packages so designing a 4-stage oscillator with 45° of phase shift relative to each other is relatively easy.

RC Oscillators are stable and provide a well-shaped sine wave output with the frequency being proportional to 1/RC and therefore, a wider frequency range is possible when using a variable capacitor. However, RC Oscillators are restricted to frequency applications because of their bandwidth limitations to produce the desired phase shift at high frequencies.

Determine the frequency of oscillations of an **RC Oscillator** circuit having 3-stages each with a resistor and capacitor of equal values. R = 10kΩ and C = 500pF. The frequency of oscillations for an RC Oscillator is given as:

$$f_r = \frac{1}{2\pi RC \sqrt{2N}}$$

The circuit is a 3-stage oscillator which consists of three 10kΩ resistors and three 500pF capacitors therefore, the frequency of oscillation is given as:

$$f = \frac{1}{2\pi \sqrt{(2\times3)} \times 10000 \times 500\times10^{-12}} = 12{,}995\,\text{Hz or 13kHz}$$

Wein Bridge Oscillator

It is one of the most popular types of oscillators used in audio and sub-audio frequency ranges (20 – 20 kHz). This type of oscillator is simple in design, compact in size, and remarkably stable in its frequency output. Furthermore, its output is relatively free from distortion and its frequency can be varied easily. However, the maximum frequency output of a typical Wien bridge oscillator is only about 1 MHz. This is also, in fact, a phase-shift oscillator. It employs two transistors, each producing a phase shift of 180°, and thus producing a total phase-shift of 360° or 0°. The circuit diagram of Wien bridge oscillator is shown in the Figure 1.5.

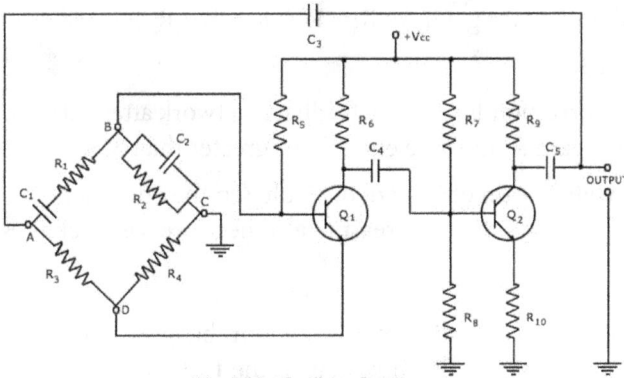

Figure 1.5: Wien Bridge Oscillator Circuit.

It is essentially a two-stage amplifier with an R-C bridge circuit. R-C bridge circuit (Wien bridge) is a lead-lag network. The phase'-shift across the network lags with increasing frequency and leads to decreasing frequency. By adding Wien-bridge feedback network, the oscillator becomes sensitive to a signal of only one particular frequency. This particular frequency is that at which Wien bridge is balanced and for which the phase shift is 0°. If the Wien-bridge feedback network is not employed and output of transistor Q_2 is fed back to transistor Q_1 for providing regeneration required for producing oscillations, the transistor Q_1 will amplify signals over a wide range of frequencies and thus direct coupling would result in poor frequency stability. Thus by employing Wien-bridge feedback network frequency stability is increased.

In the bridge circuit R_1 in series with C_1, R_3, R_4 and R_2 in parallel with C_2 form the four arms.

This bridge circuit can be used as feedback network for an oscillator, provided that the phase shift through the amplifier is zero. This requisite condition is achieved by using a two-stage amplifier, as illustrated in the figure. In this arrangement, the output of the second stage is supplied back to the feedback network and the voltage across the parallel combination $C_2 R_2$ is fed to the input of the first stage. Transistor Q_1 serves as an oscillator and amplifier whereas the transistor Q_2 as an inverter to cause a phase shift of 180°. The circuit uses positive and negative feedbacks. The positive feedback is through R_1 C_1 R_2, C_2 to transistor Q_1 and negative feedback is through the voltage divider to the input of transistor Q_1. Resistors R_3 and R_4 are used to stabilize the amplitude of the output.

The two transistors Q_1 and Q_2 thus cause a total phase shift of 360° and ensure proper positive feedback. The negative feedback is provided in the circuit to ensure constant output over a range of frequencies. This is achieved by taking resistor R_4 in the form of a temperature sensitive lamp, whose resistance increases with the increase in current. In case the amplitude of the output tends to increase, more current would provide more negative feedback. Thus the output would regain its original value. A reverse action would take place in case the output tends to fall.

The amplifier voltage gain, A $R_3 + R_4 / R_4 = R_3 / R_4 + 1 = 3$

Since $R_3 = 2 R_4$

The above corresponds with the feedback network attenuation of 1/3. Thus, in this case, voltage gain A, must be equal to or greater than 3, to sustain oscillations.

To have a voltage gain of 3 is not difficult. On the other hand, to have a gain as low as 3 may be difficult. For this reason, also negative feedback is essential.

Operation

The circuit is set in oscillation by any random change in base current of transistor Q_1, which may be due to noise inherent in the transistor or variation in voltage of dc supply. This variation in base current is amplified in the collector circuit of transistor Q_1 but with a phase-shift of 180°. The output of transistor Q_1 is fed to the base of second transistor Q_2 through capacitor C_4. Now a still further amplified and twice phase-reversed signal appears at the collector of the transistor Q_2. Having been inverted twice, the output signal will be in phase with the signal input to the base of transistor Q_1 A part of the output signal at transistor Q_2 is feedback to the input points of the bridge circuit (point A-C). A part of this feedback signal is applied to emitter resistor R_4 where it produces a degenerative effect (or negative feedback). Similarly, a part of the feedback signal is applied across the base-bias resistor R_2 where it produces the regenerative effect (or positive feedback). At the rated frequency, the effect of regeneration is made slightly more than that of degeneration so as to obtain sustained oscillations.

The continuous frequency variation in this oscillator can be had by varying the two capacitors C_1 and C_2 simultaneously. These capacitors are variable air-gang capacitors. We can change the frequency range of the oscillator by switching into the circuit different values of resistors R_1 and R_2.

Advantages

☆ Provides a stable low distortion sinusoidal output over a wide range of frequency.

☆ The frequency range can be selected simply by using decade resistance boxes.

☆ The frequency of oscillation can be easily varied by varying capacitances C_1 and C_2 simultaneously. The overall gain is high because of two transistors.

Disadvantages

☆ The circuit needs two transistors and a large number of other components.

☆ The maximum frequency output is limited because of amplitude and the phase-shift characteristics of the amplifier.

Transistor Crystal Oscillator

Introduction

Crystal oscillators are used in a variety of applications. In some instances, crystal oscillators may be used to provide a cheap clock signal for use in a digital or logic circuit. In other instances, they may be used to provide an RF signal source. In view of the fact that quartz crystals offer a very high level of Q and they are stable, crystal oscillators are often used in oscillator circuits to provide stable, accurate radio frequency signals.

Colpitts Crystal Oscillator

There is a great number of different types of circuit that can be used for crystal oscillators, each one having its own advantages and disadvantages. One of the most common circuits used for crystal oscillators is the Colpitts configuration as shown in Figure 1.6.

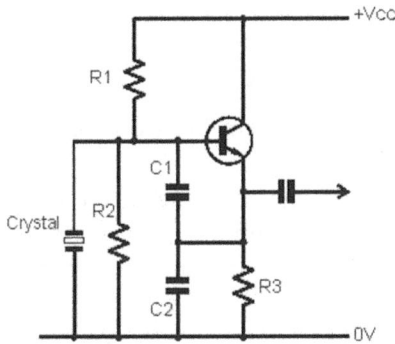

Figure 1.6: Typical Colpitts Transistor Crystal Oscillator Circuit.

The circuit uses a capacitor divider network comprising C1 and C2 to provide the feedback and the output is taken either from the emitter as shown. Alternatively, it is possible to place a resistor or choke in the collector circuit and take the output from there. In either case, it is wise to employ a buffer after the crystal oscillator circuit to ensure the minimum load is applied.

In this configuration, the crystal operates in a parallel mode. When running in this mode, the crystal should be presented with a load capacitance to operate at its correct frequency. This load capacitance is specified with the crystal and is typically 20 or 30 pF. The crystal oscillator circuit will be designed to present this capacitance to the crystal. Most of this will be made up by the two capacitors C1 and C2, although the remaining elements of the circuit will provide some capacitance.

The disadvantage with this circuit is that the resistor bias chain shunts the series combination of C1 and C2 as well as the crystal. This means that additional gain and current are required in the crystal oscillator circuit to overcome this, and also the stability may be affected to some degree. The other effect the bias resistors have is to reduce the Q of the crystal. The problem can be overcome to some degree by using a field effect transistor for the active device, but these devices are generally not as stable as bipolar devices and they often need a higher operating current.

Often a small trimmer capacitor can be used to finely adjust the frequency of the crystal oscillator to compensate for any inaccuracies, the circuit conditions, and any aging. If a parallel capacitor is used, this should enable the crystal, operating in its parallel mode to have the correct load capacitance, but care should be taken not to excessively load the crystal as this could affect the Q of the tuned circuit and reduce the performance of some applications in terms of phase noise or stability. If these parameters are critical, it may be necessary to opt for a series capacitor, although care has to be taken to ensure the value does not go too low where the oscillation will cease. Often a fixed value capacitor, sufficient to maintain reliable oscillation, will be added in parallel with the series variable capacitor in these instances.

Crystal Oscillator Gain and Drive Level

In order to obtain the best performance from a crystal oscillator, it is necessary to ensure the crystal is driven at the correct level. If the drive level for the crystal is too high then the parasitic resonances of the crystal may be excited. Alternatively, the crystal oscillator may even run on the incorrect frequency. Additionally, if the drive level is too high then the phase noise performance of the crystal oscillator will be degraded.

Additionally, the crystal can be damaged if the drive level is too high. In particular, the miniature types are susceptible to damage. Even if permanent damage is not caused, the high level of drive within the crystal oscillator increases the rate of aging and can cause a frequency shift. It is therefore important to ensure the level of drive within the crystal oscillator circuit is approximately correct.

In view of the need to ensure the correct operating conditions for the crystal oscillator itself, it is necessary to optimize the circuit for stability, gain and drive level. This may lead to a lower output level, but this can be overcome in the following stages.

Crystal Oscillator Component Value Optimization

The circuit conditions are fundamentally governed by capacitors C1, and C2 along with the bias resistors R1 and R2, and the emitter resistor R3. As the circuit is frequency dependent the values will change according to the frequency of operation. Typical values are given below.

FREQUENCY RANGE	C1 PF	C2 PF	R1 KOHMS	R2 KOHMS	R3 KOHMS
1 – 3	220	330	33	33	6.8
3 – 6	150	220	33	33	6.8
6 – 10	150	220	33	33	4.7
10 – 20	100	150	33	33	2.2

These values will give provide a good solution for many circumstances. The transistor can be a BC109 or similar general purpose transistor. The transistor crystal oscillator circuit described provides a good stable reference signal that will be satisfactory for many applications. In some circumstances, highly stable oscillators will be needed and it may be necessary to use a purpose designed and made oven controlled crystal oscillator (OCXO). These are considerably more expensive, but offer very high levels of performance in terms of stability, frequency accuracy, and phase noise. If these are needed then the additional cost may be justified.

Power Amplifier

Introduction

Amplifier circuits form the basis of most electronic systems, many of which need to produce high power to drive some output device. Audio amplifier output power may be anything from less than 1 Watt to several hundred Watts. Radio frequency amplifiers used in transmitters can be required to produce thousands of kilowatts of output power, and DC amplifiers used in electronic control systems may also need high power outputs to drive motors or actuators of many different types. This module describes some commonly encountered classes of power output circuits and techniques used to improve performance.

A power amplifier is an electronic amplifier designed to increase the magnitude of power of a given input signal. The power of the input signal is increased to a level high enough to drive loads of output devices like speakers, headphones, RF transmitters *etc*. Unlike voltage/current amplifiers, a power amplifier is designed to drive loads directly and is used as a final block in an amplifier chain.

The input signal to a power amplifier needs to be above a certain threshold. So instead of directly passing the raw audio/RF signal to the power amplifier, it is first pre-amplified using current/voltage amplifiers and is sent as input to the power amp after making necessary modifications. You can observe the block diagram of an audio amplifier and the usage of power amplifier in Figure 2.1.

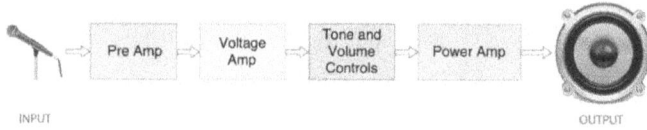

Figure 2.1: Block Diagram of an Audio Amplifier.

The magnitude of the signal from the microphone is not enough for the power amplifier. So first it is pre-amplified where its voltage and current are increased slightly. Then the signal is passed through tone and volume controls circuit which makes aesthetic adjustments to the audio waveform. Finally, the signal is passed through a power amplifier and the output from power amp is fed to a speaker.

Types of Power Amplifiers

Depending on the type of output device that is connected, power amplifiers are divided into the following three types:

Audio Power Amplifiers

This type of power amplifiers is used for increasing the magnitude of power of a weaker audio Signal. The amplifiers used in speaker driving circuitries of televisions, mobile phones *etc.* come under this category. The output of an audio power amplifier ranges from a few milliwatts (like in headphone amplifiers) to thousands of watts (like power amplifiers in Hi-Fi/Home theatre systems).

Radio Frequency Power Amplifiers

Wireless transmissions require modulated waves to be sent over long distances via air. The signals are transmitted using antennas and the range of transmission depends on the magnitude of power of signals fed to the antenna.

For wireless transmissions like FM broadcasting, antennas require input signals at thousands of kilowatts of power. Here, Radio Frequency Power amplifiers are employed to increase the magnitude of power of modulated waves to a level high enough for reaching required transmission distance.

DC Power Amplifiers

DC power amplifiers are used to amplify the power of a PWM (Pulse Width Modulated) signals. They are used in electronic control systems which need high power signals to drive motors or actuators. They take input from microcontroller systems, increase its power and feed the amplified signal to DC motors or Actuators.

Power Amplifier Classes

There are multiple ways of designing a power amplifier circuit. The operation and output characteristics of each of the circuit configurations differ from each other. To differentiate the characteristics and behavior of different power amplifier circuits, Power Amplifier Classes are used in which letter symbols are assigned to identify the method of operation.

They are broadly classified into two categories. Power amplifiers designed to amplify analog signals come under A, B, AB or C category. Power amplifiers designed to amplify Pulse Width Modulated(PWM) digital signals come under D, E, F *etc.* The most commonly used power amplifiers are the ones that are used in audio amplifier circuits and they come under classes A, B, AB or C.

Class A Power Amplifier

If the collector current flows all at all times during the full cycle of the signal, the power amplifier is known as class A power amplifier. To achieve this, the power amplifier must be biased in such a way that no part of the signal is cut off. The circuit diagram of a class A power amplifier is shown in Figure 2.2.

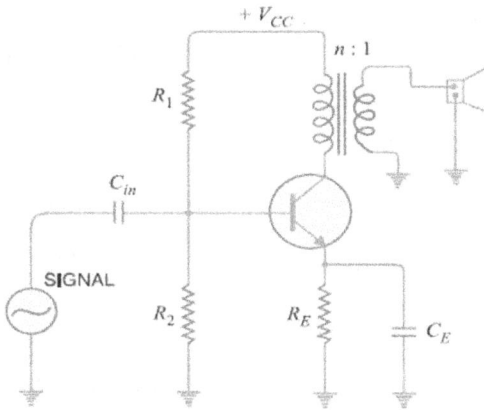

Figure 2.2: Class A Power Amplifier.

In case of a direct-coupled class A power amplifier, as shown above, the current flows through the collector resistive load causes large wastage of dc power in it. As a result, this dc power dissipated in the load resistor does not contribute to the useful ac output power.

Hence, it is generally inadvisable to pass the current through the output device such as in a voice coil of a loudspeaker. For these reasons an arrangement using a suitable transformer for coupling the load to the amplifier is usually employed, as shown above. This arrangement also permits impedance matching.

In a power amplifier circuit shown R1 and R2 provide potential divider biasing and emitter resistor RE is meant for bias stabilization. The emitter bypass capacitor CE is meant for RE to prevent ac voltage. The input capacitor Cin couples ac signal voltage to the base of the transistor but blocks any dc from the previous stage. A step-down transformer of suitable turn ratio is provided to couple the high impedance collector circuit to low impedance load.

The power transferred from the power amplifier to the load such as loudspeaker will be maximum only if the amplifier output impedance equals the load impedance RL. This is in accordance with the maximum power transfer theorem.

If we were not able to achieve the above condition, lesser power will be transferred to the load RL, though the amplifier is capable of delivering more power, and rest of power developed would be lost in the active device. Hence for transfer of maximum power from the amplifier to the output device matching of amplifier output impedance with the impedance of output device is necessary. This is accomplished by using a step-down transformer of suitable turn-ratio. The operation of a class A power amplifier as A.C. load line is shown in Figure 2.3.

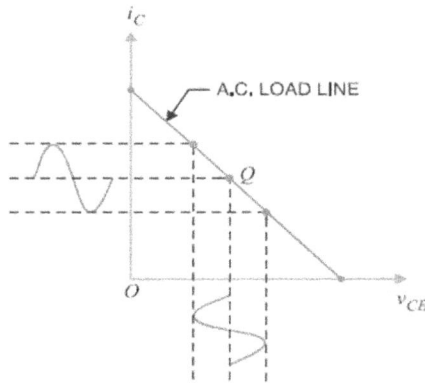

Figure 2.3: A.C. Load Line of Class A Amplifier.

The operating point Q is so selected that collector current flows at all times throughout the full cycle of the applied signal. Since the output wave shape is exactly similar to the input wave shape, hence, such amplifiers have the least distortion. The only disadvantage of class A power amplifier is the low output power and low collector efficiency.

Class B Power Amplifier

If the collector current flows only during the positive half-cycle of the input signal, it is called as class B power amplifier. In class B power amplifier operation, the transistor is so adjusted that zero signal collector current is zero *i.e.* no biasing circuit i needed at all.

During the positive half-cycle of the signal, the input circuit is forward biased and hence collector current flows. During the negative half-cycle, the input circuit is reversing biased and no collector current flows. The operation of the class B power amplifier in terms of A.C. load line is shown in Figure 2.4.

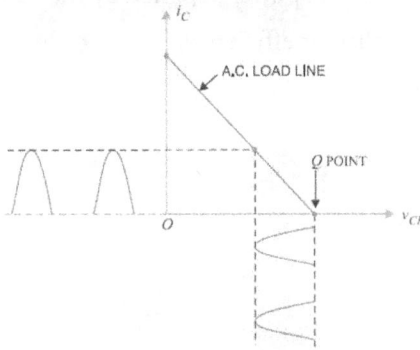

Figure 2.4: A.C. Load Line of Class B Amplifier.

As you can see, the operating point Q is located at collector cut off voltage. In class B amplifier, the negative half-cycle of the signal is cut off and hence severe distortion occurs. However, it provides higher power output and hence collector efficiency.

These amplifiers are mostly used for power amplification in a push-pull arrangement. In such arrangement, two transistors are used in class B operation. One transistor amplifies the positive half cycle of the signal and the other one amplifies the negative half-cycle of the signal.

Class C Power Amplifier

If the collector current flows for less than half-cycle of the input signal, it is called class C power amplifier. In class C power amplifier, the base is negatively biased, so that collector current does not flow just when the positive half-cycle of the signal starts.

Such amplifiers are never used for power amplification but as tuned amplifier *i.e.* to amplify a narrow band of frequencies near the resonant frequency. The performance quantities or criteria for a power amplifier are:

☆ Collector Efficiency

☆ Distortion

☆ Power Dissipation Capability

(i) Collector Efficiency

The main criterion for a power amplifier is not the power gain but the maximum A.C. power output. An amplifier converts D.C. power from supply into A.C. power output. Hence, the effectiveness of a power amplifier is measured in terms of its ability to convert D.C. power from the supply to A.C. output power. This is called as collector efficiency. The collector efficiency is defined as the ratio of output power to the zero signal power or D.C. power supplied by the battery.

The expression for collector efficiency

Collector efficiency,

$$\eta = \frac{A..C.\text{power output}}{D.C.\text{power output}}$$

$$= \frac{P_o}{P_{dc}}$$

Where,

$$P_{dc} = V_{cc}I_c$$
$$P_o = V_{ce}I_c$$

Where Vce is the r.m.s. value of signal output voltage and Ic is the r.m.s. value of output signal current. In terms of peak-to-peak values, the A.C. power output can be expressed as :

$$P_o = [(0.5\times0.707)v_{ce\,(p-p)}][(0.5\times0.707)i_{c\,(p-p)}]$$
$$= \frac{v_{ce(p-p)}\times i_{c(p-p)}}{8}$$

$$\therefore \quad \text{Collector } \eta = \frac{v_{ce\,(p-p)}\times i_{c(p-p)}}{8V_{cc}I_c}$$

(ii) Distortion

The change of output wave shape from the input wave shape of an amplifier is known as distortion. Since the transistor is a non-linear device, the output is not exactly like the input signal applied to it, which lead to distortion occurs.

(iii) Power Dissipation Capability

The ability of a power transistor to dissipate heat is known as power dissipation capability. As we know a power transistor handles large currents and hence, heats up during operation. Since, any temperature change influences the operation of the transistor; therefore, the transistor must dissipate this heat to its surroundings. To achieve this, normally a heat sink is attached to a power transistor case. The increased surface area allows heat to escape easily and keeps the case temperature of the transistor within permissible limit.

Class D Amplifiers

In class D audio amplifiers, the basic operation of which is shown in Figure 2.5, the audio signal is first converted to a type of digital signal called 'Pulse Width Modulation'. This is not a digital signal within the normally accepted definition of 'Digital' regarding recognized logic 1 and 0 levels, but only in that, it has two levels, high and low. When such a signal is amplified, very little power is dissipated in the amplifier, as when the output transistors are 'on' and passing their maximum current, there is practically no voltage drop across them and therefore (as Power = Voltage × Current) there is practically zero Power. Likewise, when the transistors are 'off', a large voltage is present but no current is flowing. This results in much greater efficiency than in conventional analogue amplifiers. The PWM output signal is finally converted back into analog form at the output.

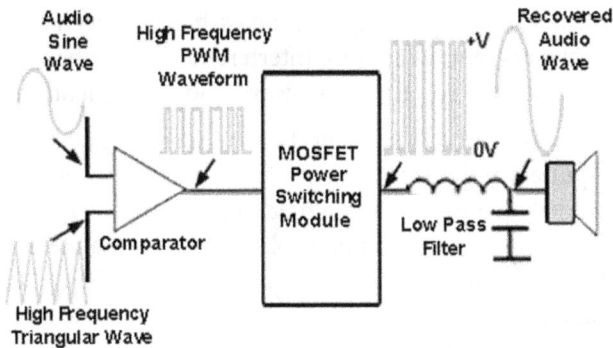

Figure 2.5: Principles of Class D.

Fast Power Switching

This PWM output then drives a switching circuit that uses fast switching MOSFET power transistors to switch the output between the full supply voltage (+V) and zero volts (0V). Representing the audio by a series of square pulses dramatically reduces power consumption. When the square wave is at its high level, the transistor is cut-off and although a high voltage is present, virtually no current is flowing, and as power is voltage × current, power is just about zero. During the time when the square wave signal is at its low level, there will be a large current flowing but the signal voltage is practically zero, so again, very little power dissipation. The only time when significant power is dissipated by the MOSFET transistors is during the change from high to low or low to high states. Because high speed switching MOSFETs are used, this period is extremely short, so average power dissipation is kept to a very low level.

Output Filtering

Finally the high frequency, large amplitude PWM pulses are applied to a low pass filter, which removes the high frequency components of the waveform, leaving just the average level of the PWM wave, which because of the varying 'on' time of the pulses produces a large amplitude replica of the original sine wave audio input.

Class D operation makes the output circuit extremely efficient (around 90%) allowing high power output without the need for such high power transistors and elaborate heat-sinks. However ,this big increase in efficiency is only achieved at the expense of some increase in distortion and especially of noise, in the form of electromagnetic interference (EMI).

Because class D produces high frequency, large amplitude PWM 'square' waves at the output of the switching module, there will be many large amplitude odd harmonics present, which are a natural component of square waves. The frequencies of these harmonics can extend well into the radio frequency spectrum, and if not carefully controlled can cause radio interference radiated both directly, and conducted via the power supply.

As shown in above fig, PWM output is fed to the load via a low pass filter, which should remove the interference causing harmonics, however even with a good quality low pass filter all the harmonics causing interference cannot be removed from the square wave, as this would distort the square wave, and consequently the audio output.

Nevertheless, class D is a very efficient class of amplifier suited to both high power audio and RF amplifiers and low power portable amplifiers, where battery life can be considerably extended because of the amplifier's high efficiency. The increased interest in class D amplifiers has led to a number of class D integrated circuits becoming available, such as the 25 watts per channel stereo power amplifier TPA3120D2 from Texas Instruments and the 2 watts per channel digital input audio amplifier SSM2518 from Analogue Devices, suitable for mobile phones and portable mp3 players.

Classes E to H

Amplifier classes such as E and F are basically enhancements of class D, offering more complex and improved output filtering, including some additional wave shaping of the PWM signal to prevent audio distortion.

Classes G and H offer enhancements to the basic class AB design. Class G uses multiple power supply rails of various voltages, rapidly switching to a higher voltage when the audio signal wave has a peak value that is a higher voltage than the level of supply voltage and switching back to a lower supply voltage when the peak value of the audio signal reduces. By switching the supply voltage to a higher level only when the largest output signals are present and then switching back to a lower level, average power consumption, and therefore heat caused by wasted power is reduced.

Class H improves on class G by continually varying the supply voltage at any time where the audio signal exceeds a particular threshold level. The power supply voltage tracks the peak level of the signal to be only slightly higher than the instantaneous value of the audio wave, returning to its lower level once the signal peak value falls below the threshold level again. Both classes G and H therefore, require considerably more complex power supplies, which add to the cost of implementing these features.

BJT and Field Effect Transistor

BJT

BJT is bipolar junction transistor, which is very versatile and can be used in many ways, as an amplifier, a switch or an oscillator and many other uses too. Before an input signal is applied its operating conditions need to be set. This is achieved with a suitable bias circuit, some of which I will describe. A bias circuit allows the operating conditions of a transistor to be defined so that it will operate over a predetermined range. This is normally achieved by applying a small fixed DC voltage to the input terminals of a transistor.

Bias design can take a mathematical approach or can be simplified using transistor characteristic curves. The characteristic curves predict the performance of a BJT. There are three curves, an input characteristic curve, a transfer characteristic curve and an output characteristic curve. Of these curves, the most useful for amplifier design is the output characteristics curve. The output characteristic curves for a BJT are a graph displaying the output voltages and currents for different input currents. The linear (straight) part of the curve needs is utilized for an amplifier or oscillator. For use as a switch, a transistor is biased at the extremities of the graph, these conditions are known as "cut-off" and "saturation".

Types of BJT's

Transistors are three- terminal active devices made from different semiconductor materials that can act as either an insulator or a conductor by the application of a small signal voltage. The transistor's ability to change between these two states enables it to have two basic functions: "switching" (digital electronics) or "amplification" (analogue electronics). Then bipolar transistors have the ability to operate within three different regions:

1. Active Region - the transistor operates as an amplifier and Ic = β.Ib
2. Saturation - the transistor is «fully-ON» operating as a switch and Ic = I (saturation)
3. Cut-off - the transistor is «fully-OFF» operating as a switch and Ic = 0

The word Transistor is an acronym, and is a combination of the words Transfer Varistor used to describe their mode of operation way back in their early days of development. There are two basic types of bipolar transistor construction, PNP, and NPN, which basically describes the physical arrangement of the P-type and N-type semiconductor materials from which they are made.

The Bipolar Transistor basic construction consists of two PN-junctions producing three connecting terminals with each terminal being given a name to identify it from the other two. These three terminals are known and labeled as the Emitter (E), the Base (B) and the Collector (C) respectively. Bipolar Transistors are current regulating devices that control the amount of current flowing through them in proportion to the amount of biasing voltage applied to their base terminal acting like a current-controlled switch. The principle of operation of the two transistor types PNP and NPN is exactly the same the only difference being in their biasing and the polarity of the power supply for each type (Figure 3.1).

Figure 3.1: Typical Bipolar Transistor.

Construction

The construction and circuit symbols for both the PNP and NPN bipolar transistor are given below with the arrow in the circuit symbol always showing the direction of "conventional current flow" between the base terminal and its emitter terminal. The direction of the arrow always points from the positive P-type region to the negative N-type region for both transistor types, exactly the same as for the standard diode symbol (Figure 3.2).

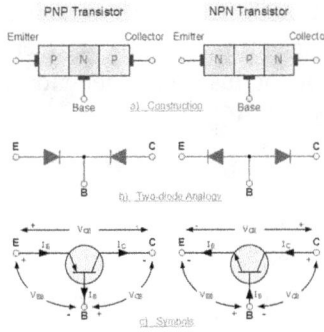

Figure 3.2: BJT Construction.

Transistor Terminals

The essential usefulness of a transistor comes from its ability to use a small signal applied between one pair of its terminals to control a much larger signal at another pair of terminals. This property is called gain. A transistor can control its output in proportion to the input signal; that is, it can act as an amplifier. Alternatively, the transistor can be used to turn current on or off in a circuit as an electrically controlled switch, where the amount of current is determined by other circuit elements.

There are two types of transistors, which have slight differences in how they are used in a circuit. A *bipolar transistor* has terminals labeled **base, collector,** and **emitter**. A small current at the base terminal (that is, flowing from the base to the emitter) can control or switch a much larger current between the collector and emitter terminals. For a *field-effect transistor*, the terminals are labeled **gate, source,** and **drain**, and a voltage at the gate can control a current between source and drain.

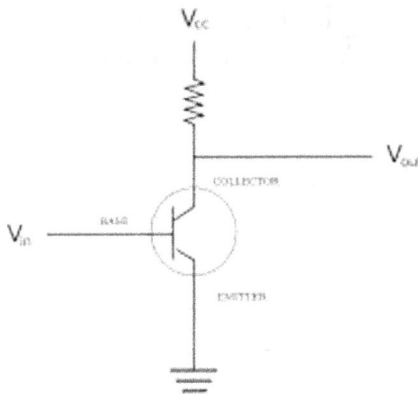

Figure 3.3: Transistor.

The image to the right represents a typical bipolar transistor in a circuit. The charge will flow between emitter and collector terminals depending on the current in the base. Since internally the base and emitter connections behave like a semiconductor diode, a voltage drop develops between base and emitter while the base current exists. The amount of this voltage depends on the material the transistor is made from and is referred to as V_{BE}.

Transistor Action

Bipolar transistors are made from 3 sections of semiconductor material (alternating P-type and N-type), with 2 resulting P-N junctions. Schematically, a bipolar transistor can be thought of in this fashion (Figure 3.4).

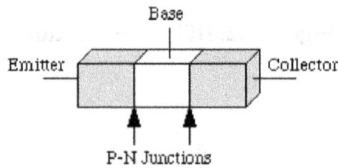

Figure 3.4: BJT.

One P-N junction is between the emitter and the base; the other P-N junction is between the collector and the base. Note that the emitter and collector are usually doped somewhat differently, so they are rarely electrically interchangeable. While the terms "collector" and "emitter" go back to vacuum tube days, the base derives its name from the first point-contact transistors—here the center connection also formed the mechanical base for the structure. In modern practice, the base region is made as thin as possible to achieve reasonable levels of current gain; it is often only about one m-illionth of a meter thick.

Bipolar transistors are classified as either NPN or PNP according to the arrangement of their N-type and P-type materials. Their basic construction and chemical treatment is implied by their names. So an NPN transistor is formed by introducing a thin region of the P-type material between two regions of the N-type material (Figure 3.5).

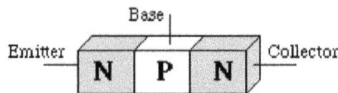

Figure 3.5: NPN Transistor.

On the other hand, a PNP transistor is formed by introducing a thin region of N-type material between two regions of P-type material (Figure 3.6).

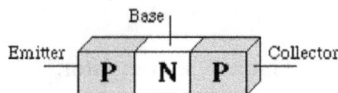

Figure 3.6: PNP Transistor.

Since the majority and minority current carriers are different for N-type and P-type materials, it stands to reason that the internal operation of the NPN and PNP transistors will also be different. These two basic types of transistors along with their circuit symbols are shown here:

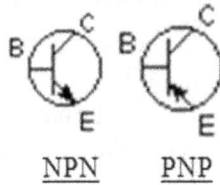

NPN   PNP

Note that the two symbols are subtly different. The vertical line represents the base (B), the angular line with the arrow on it represents the emitter (E), and the other angular line represents the collector (C). The direction of the arrow on the emitter distinguishes (graphically) the NPN from the PNP transistor. If the arrow points in, (Points iN) the transistor is a PNP. On the other hand, if the arrow points out, the transistor is an NPN (Not Pointing iN).

Bear in mind that the arrow always points in the direction of hole flow (current), or from the P-type to N-type sections, no matter whether the P-type section is the emitter or base. On the other hand, electron flow is always "against" the arrow, just like in the junction diode. As a result, a PNP transistor is "triggered" when its base is pulled low; an NPN transistor is "triggered" when its base is brought high.

Note that the bipolar transistor is a current-amplifying device, unlike the vacuum tube and the field-effect transistor (FET), both of which depend upon voltage changes to operating. It is the amount of current flowing in the base circuit that controls the amount of current flowing in the collector circuit.

Transistor Configurations/Connections

You have to think in terms of circuit configurations and the voltage and current in each lead when discussing how transistors behave. There are 3 configurations—the emitter follower which is a current amplifier but has no voltage gain, the common emitter amplifier which has current and voltage gain, and the common base amplifier which has voltage gain but no current gain.

In an emitter follower circuit with the collector connected to +V and a load connected between the emitter and ground, the voltage applied to the base minus the base- emitter forward voltage drop (~0.6 V) will appear across the load (*i.e.*, 5 V base = 4.4 V emitter). The only caveat is that the voltage source at the base must be able to supply about 5% of the load current without appreciable voltage drop. This is a non-inverting voltage follower circuit.

In a common emitter circuit with the emitter connected to ground and the load connected between the collector and +V, a voltage connected to the base which exceeds the base- emitter forward voltage (0.6V) will rapidly turn on the transistor in proportion to the voltage rise as the base- emitter current rapidly increases for a small increase in base voltage. The base voltage source must be able to supply about 5% of the load current into the base emitter diode (*i.e.*, short circuit) for the circuit to develop a large voltage across the load. This is an inverting voltage amplifier circuit.

In a common base circuit with the base grounded (or at a reference voltage) and the load connected between the collector and +V, a control voltage connected to the emitter which is more negative than the base- emitter forward voltage (~0.6V) causes the transistor to rapidly turn on. The control voltage source must be able to supply about 105% of the load current to develop the full voltage across the load. This is a non-inverting voltage amplifier circuit

Common-Base (CB) Configuration

As its name suggests, in the **Common Base** or grounded base configuration, the BASE connection is common to both the input signal AND the output signal with the input signal being applied between the base and the emitter terminals. The corresponding output signal is taken from between the base and the collector terminals as shown with the base terminal grounded or connected to a fixed reference voltage point. The input current flowing into the emitter is quite large as its the sum of both the base current and collector current respectively, therefore, the collector current output is less than the emitter current input resulting in a current gain for this type of circuit of "1" (unity) or less, in other words, the common base configuration "attenuates" the input signal (Figure 3.7).

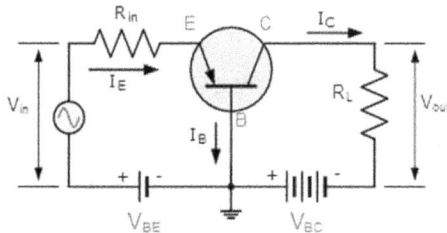

Figure 3.7: CB Transistor.

This type of amplifier configuration is a non-inverting voltage amplifier circuit, in that the signal voltages Vin and Vout are in-phase. This type of transistor arrangement is not very common due to its unusually high voltage gain characteristics. Its output characteristics represent that of a forward biased diode while the input characteristics represent that of an illuminated photo-diode. Also, this type of bipolar transistor configuration has a high ratio of output to input resistance or more importantly "load" resistance (RL) to "input" resistance (Rin) giving it a value of "Resistance Gain". Then the voltage gain (Av) for a common base configuration is therefore given as:

$$A_V = \frac{Vout}{Vin} = \frac{I_C \times R_L}{I_E \times R_{IN}}$$

Where: Ic/Ie is the current gain, alpha (α) and RL/Rin is the resistance gain.

The common base circuit is generally only used in single stage amplifier circuits such as microphone pre-amplifier or radio frequency (Rf) amplifiers due to its very good high- frequency response.

Common-Emitter (CE) Configuration

In the **Common Emitter** or grounded emitter configuration, the input signal is applied between the base, while the output is taken from between the collector and the emitter as shown in Figure 3.8. This type of configuration is the most commonly used circuit for transistor- based amplifiers and which represents the "normal" method of bipolar transistor connection. The common emitter amplifier configuration produces the highest current and power gain of all the three bipolar transistor configurations. This is mainly because the input impedance is LOW as it is connected to a forward-biased PN-junction, while the output impedance is HIGH as it is taken from a reverse-biased PN-junction.

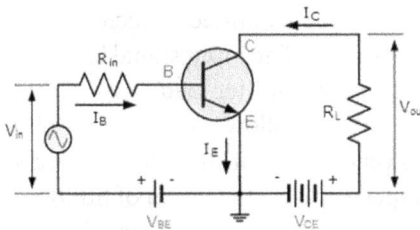

Figure 3.8: CE Transistor.

In this type of configuration, the current flowing out of the transistor must be equal to the currents flowing into the transistor as the emitter current is given as Ie = Ic + Ib. Also, as the load resistance (RL) is connected in series with the collector, the current gain of the common emitter transistor configuration is quite large as it is the ratio of Ic/Ib and is given the Greek symbol of Beta, (β). As the emitter current for a common emitter configuration is defined as Ie = Ic + Ib, the ratio of Ic/Ie is called Alpha, given the Greek symbol of α. Note: that the value of Alpha will always be less than unity.

Since the electrical relationship between these three currents, Ib, Ic and Ie is determined by the physical construction of the transistor itself, any small change in the base current (Ib), will result in a much larger change in the collector current (Ic). Then, small changes in current flowing in the base will thus control the current in the emitter-collector circuit. Typically, Beta has a value between 20 and 200 for most general purpose transistors. By combining the expressions for both Alpha, α

and Beta, β the mathematical relationship between these parameters and therefore the current gain of the transistor can be given as:

$$\text{Alpha}, (\alpha) = \frac{I_C}{I_E} \quad \text{and} \quad \text{Beta}, (\beta) = \frac{I_C}{I_B}$$

$$\therefore I_C = \alpha.I_E = \beta.I_B$$

$$\text{as: } \alpha = \frac{\beta}{\beta + 1} \qquad \beta = \frac{\alpha}{1 - \alpha}$$

$$I_E = I_C + I_B$$

Where: "Ic" is the current flowing into the collector terminal, "Ib" is the current flowing into the base terminal and "Ie" is the current flowing out of the emitter terminal. Then to summarize, this type of bipolar transistor configuration has a greater input impedance, current, and power gain than that of the common base configuration but its voltage gain is much lower. The common emitter configuration is an inverting amplifier circuit resulting in the output signal being 180°out-of-phase with the input voltage signal.

Common Collector (CC) Configuration

In the Common Collector or grounded collector configuration, the collector is now common through the supply. The input signal is connected directly to the base, while the output is taken from the emitter load as shown. This type of configuration is commonly known as a Voltage Follower or Emitter Follower circuit. The emitter follower configuration is very useful for impedance matching applications because of the very high input impedance, in the region of hundreds of thousands of Ohms while having a relatively low output impedance. The common collector (CC) configuration is depicted in Figure 3.9.

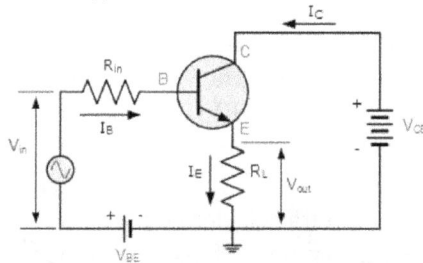

Figure 3.9: CC Transistor Configuration.

The common emitter configuration has a current gain approximately equal to the β value of the transistor itself. In the common collector configuration, the load resistance is situated in series with the emitter so its current is equal to that of the emitter current. As the emitter current is the combination of the collector AND the base current combined, the load resistance in this type of transistor configuration also has both the collector current and the input current of the base flowing through it. Then the current gain of the circuit is given as:

$$I_E = I_C + I_B$$

$$A_i = \frac{I_E}{I_B} = \frac{I_C + I_B}{I_B}$$

$$A_i = \frac{I_C}{I_B} + 1$$

$$A_i = \beta + 1$$

This type of bipolar transistor configuration is a non-inverting circuit in that the signal voltages of Vin and Vout are in-phase. It has a voltage gain that is always less than "1" (unity). The load resistance of the common collector transistor receives both the base and collector currents giving a large current gain (as with the common emitter configuration) therefore, providing good current amplification with very little voltage gain.

Comparison Of Three Configurations

To summarize, the behavior of the bipolar transistor in each one of the above circuit configurations is very different and produces different circuit characteristics with regards to input impedance, output impedance and gain whether this is voltage gain, current gain or power gain and this is summarized in the table below. The static characteristics for a Bipolar Transistor can be divided into the following three main groups.

Input Characteristics:- Common Base - $\Delta VEB / \Delta IE$

Common Emitter - $\Delta VBE / \Delta IB$

Output Characteristics:- Common Base - $\Delta VC / \Delta IC$

Common Emitter - $\Delta VC / \Delta IC$

Transfer Characteristics:- Common Base - $\Delta IC / \Delta IE$

Common Emitter - $\Delta IC / \Delta IB$

With the characteristics of the different transistor configurations given in the Table 3.1:

Table 3.1: Characteristics of the different Transistor Configurations

Characteristic	Common Base	Common Emitter	Common Collector
Input Impedance	Low	Medium	High
Output Impedance	Very High	High	Low
Phase Angle	0°	180°	0°

Characteristic	Common Base	Common Emitter	Common Collector
Voltage Gain	High	Medium	Low
Current Gain	Low	Medium	High
Power Gain	Low	Very High	Medium

Operating Point (Quiescent, Q Or Silent Point)

The point Vo in the diagram above is where the output signal would be taken. For simplicity, the input signal and coupling capacitors have been omitted. For minimum distortion and clipping, it is desirable to bias this point to half the supply voltage, 10 volts dc in this case. This is also known as the quiescent point. The ac output signal would then be superimposed on the dc bias voltage. The Q-point is sometimes indicated on the output characteristics curves for a transistor amplifier. The quiescent point also refers to the dc conditions (bias conditions) of a circuit without an input- signal.

Q-Point Value

I have mentioned that setting the Q-point to half the supply voltage is a good idea. It gives a circuit the highest margin for overload. However, any amplifier will clip if the input amplitude exceeds the limit for which the circuit was designed. However, there are certain cases when it is not necessary to bias a stage to half the supply voltage. Examples would be an RF amplifier design where the input signal is in microvolts or millivolts. If the stage had a gain of 200 then the output (assuming a 2mV peak input) would only need to swing up and down 400mV about the Q-point. Hence a stage with a supply voltage of 12 volts could have its Q-point set at 10 volts or even 2 volts without problems. Another example would be a microphone stage where similar low level input signals are involved.

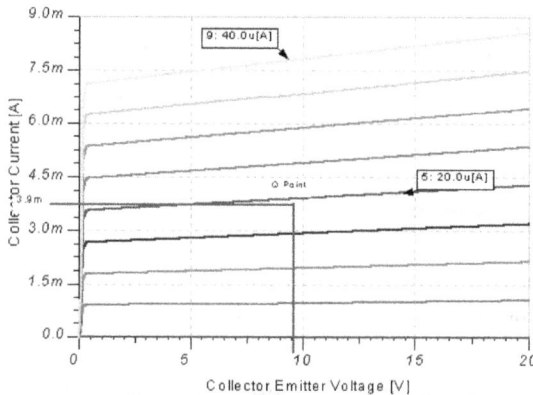

Different Operating Conditions of Transistor

Introduction

When a transistor is in the fully-off state (like an open switch), it is said to be *cutoff*. Conversely, when it is fully conductive between emitter and collector (passing as much current through the collector as the collector power supply and load will allow), it is said to be *saturated*. These are the two modes of operation explored thus far in using the transistor as a switch.

However, bipolar transistors don't have to be restricted to these two extreme modes of operation. As we learned in the previous section, the base current "opens a gate" for a limited amount of current through the collector. If this limit for the controlled current is greater than zero but less than the maximum allowed by the power supply and load circuit, the transistor will "throttle" the collector current in a mode somewhere between cutoff and saturation. This mode of operation is called the *active* mode.

An automotive analogy for transistor operation is as follows: *cutoff* is the condition of no motive force generated by the mechanical parts of the car to make it move. In cutoff mode, the brake is engaged (zero base current), preventing motion (collector current). *Active mode* is the automobile cruising at a constant, controlled speed (constant, controlled collector current) as dictated by the driver. *Saturation* the automobile driving up a steep hill that prevents it from going as fast as the driver wishes. In other words, a "saturated" automobile is one with the accelerator pedal pushed all the way down (base current calling for more collector current than can be provided by the power supply/load circuit). In the circuit, what happens when a transistor is in its active mode of operation.

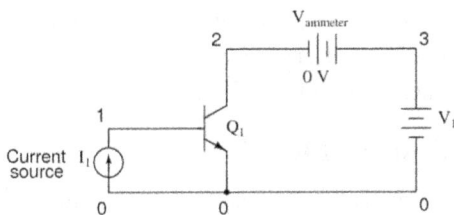

"Q" is the standard letter designation for a transistor in a schematic diagram, just as "R" is for a resistor and "C" is for a capacitor. In this circuit, we have an NPN transistor powered by a battery (V_1) and controlled by the current through a *current source* (I_1). A current source is a device that outputs a specific amount of current, generating as much or as little voltage across its terminals to ensure that exact amount of current through it. Current sources are notoriously difficult to find in nature (unlike voltage sources, which by contrast attempt to maintain a constant voltage, outputting as much or as little current in the fulfillment of that task), but can be simulated with a small collection of electronic components. As we are about to see, transistors themselves tend to mimic the constant-current behavior of a current source in their ability to *regulate* current at a fixed value.

In the circuit, we'll set the current source at a constant value of 20 µA, then vary the voltage source (V_1) over a range of 0 to 2 volts and monitor how much current goes through it. The "dummy" battery ($V_{ammeter}$) in the figure above with its output of 0 volts serves merely to provide with a circuit element for current measurement (Figure 3.10).

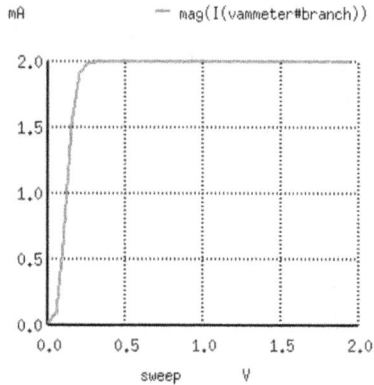

Figure 3.10: Current Measurement.

A Sweeping collector voltage 0 to 2 V with a base current constant at 20 μA yields constant 2 mA collector current in the saturation region. The constant base current of 20 μA sets a collector current limit of 2 mA, exactly 100 times as much. Notice how flat the curve is in the figure above for collector current over the range of battery voltage from 0 to 2 volts. The only exception to this featureless plot is at the very beginning, where the battery increases from 0 volts to 0.25 volts. There, the collector current increases rapidly from 0 amps to its limit of 2 mA.

Diode Resistor Transistor Model

This model casts the transistor as a combination of diode and rheostat (variable resistor). Current through the base-emitter diode controls the resistance of the collector-emitter rheostat (as implied by the dashed line connecting the two components), thus controlling collector current. An NPN transistor is modeled in the figure shown, but a PNP transistor would be only slightly different (only the base-emitter diode would be reversed). This model succeeds in illustrating the basic concept of transistor amplification: how the base current signal can exert control over the collector current. However, I don't like this model because it miscommunicates the notion of a set amount of collector-emitter resistance for a given amount of base current. If this were true, the transistor wouldn't *regulate* collector current at all like the characteristic curves show. Instead of the collector current curves flattening out after their brief rise as the collector-emitter voltage increases, the collector current would be directly proportional to collector-emitter voltage, rising steadily in a straight line on the graph. A better transistor model is shown in the Figure 3.11.

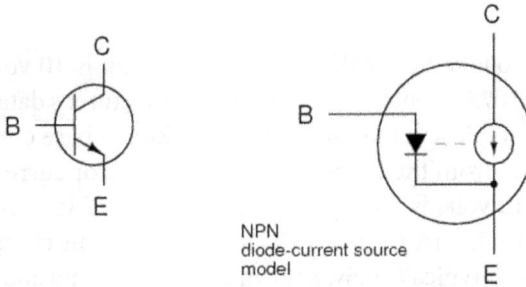

NPN
diode-current source
model

Figure 3.11: NPN Transistor Model.

Current Source Model

It casts the transistor as a combination of diode and current source, the output of the current source being set at a multiple (β ratio) of the base current. This model is far more accurate in depicting the true input/output characteristics of a transistor: base current establishes a certain amount of collector *current*, rather than a certain amount of collector-emitter *resistance* as the first model implies. Also, this model is favored when performing network analysis on transistor circuits, the current source being a well-understood theoretical component. Unfortunately, using a current source to model the transistor's current-controlling behavior can be misleading: in no way will the transistor ever act as a *source* of electrical energy. The current source does not model the fact that its source of energy is an external power supply, similar to an amplifier.

Transistor Biasing

Introduction

The simplest bias circuit is shown below. It consists only of a fixed bias resistor and load resistor. The BJT is operating in common emitter mode. The dc current gain or beta, h_{FE} is the ratio of dc collector current divided by dc base current. The BJT is a BC107A. The values of Rb and Rc can be determined by either mathematical approach or by using the output characteristic curves for the BC107A.

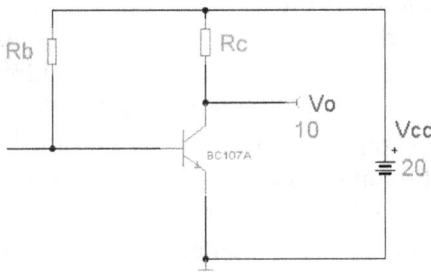

Design

The collector voltage Vc for the simple bias design is 10 volt. The dc current gain, h_{FE} for the BC107A is obtained from the manufacturer's data sheets and varies between devices. A typical beta is around 290. Taking a base current of 20uA and reading values direct from the output curves, the collector current, for a collector-emitter voltage of 10 volts is around 3.9mA. As h_{FE}= Ic / Ib then a BC107A must have a beta of at least 3.9mA / 20uA = 195 to work with this circuit. Also, the base-emitter voltage, Vbe is typically 0.6v. Knowing the above data and using Ohm's law , values for Rb and Rc can be determined:

Rb = Vcc - Vbe / Ib = (20-0.6) / 20u = 970k use (1M)

Rc = Vc / Ic = 10 / 3.9m = 2.56K use (2.7K)

Mathematical Approach

Without using the output characteristic curve, values for Rb and Rc can still be calculated. A value for h_{FE} must be estimated first and a desired collector current. As h_{FE} varies in each transistor the value chosen should be the lowest value from the manufactures data sheets. The equations to use are:

Rc = Vc / Ic

Ib = Ic / h_{FE}

Rb = Vcc - 0.6 / Ib.

Using the example above with Vcc=20 and h_{FE} =195 yields the same values.

Need for Biasing of Transistor

A BJT (Bipolar Junction Transistor) require a voltage normally in the range of 0.7V for the internal junctions to become conductive. It is a fixed parameter of Silicon (Si) due to the amount of 1.1eV required to get electrons from the valence energy band into a conductive band. To be able jump the energy gap which is a forbidden band for electrons or to raise the Fermi energy level in the atom? The energy, whether it is electrically applied, thermally or optically, is required to be able change the state of a semiconductor from an insulator to a conductor. Then with a non-linear relationship, the conductivity will increase as one increase the forward bias current through the base to emitter junction. Biasing is used for classical transistor amplifier applications. Biasing is required to have the transistor half way saturated for Class-A amplification or barely switched on for Class-B power amplifiers. If a Class-B amplifier is not biased, then the lower 0.7V of the audio or sine wave will not be amplified causing crossover distortion. When you bias it correctly, the distortions will be gone, since the entire half wave will then fit into the state of the transistor. If a Class-A amplifier is not correctly biased, premature clipping on the positive or negative part of the wave will occur. Biasing may be used for other applications as well, such as phototransistors, the internal construction of IC's such as op-amps.

This is not a problem for FM operation, where the drive level is constant, and always enough to 'turn on' the PA transistor. FM-only PAs usually have an RF choke to provide a DC return path directly from the base to ground (Figure 3.12).

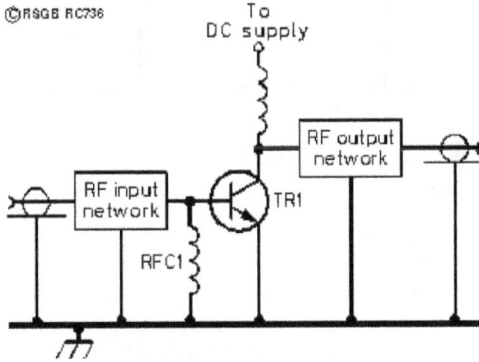

Figure 3.12: Circuit Without Bias - Only RFC1 from Base to Ground.

This is sometimes called 'Class C' operation. But for SSB modulation, the delayed turn-on at low drive levels is disastrous: it means that all the low-level parts of the modulation are severely distorted and the amplifier only delivers RF power in bursts on speech peaks. This is why you must **never** use an FM-only power amplifier on SSB - the splatter is horrendous!

For linear operation, it is **essential** to use a fixed DC bias to make sure that the transistor is already 'turned on' before any RF drive arrives (so-called 'Class AB'). Typically, if the transistor is drawing about 100mA in 'standing current' with no RF drive, distortion will be quite low, provided that the bias supply can also cope with the higher demands of peak modulation.

The standing collector current in the RF power transistor depends on the current flowing from the bias supply into the base, and on the beta of the transistor (ratio of collector current to base current). For a typical RF power transistor with a beta of 50, a standing collector current of 100mA needs a base current of 2mA. However, beta is not a very well-controlled value, so the next transistor might need only 2mA; or maybe 7mA. That's one reason why the bias supply **must** be adjustable. Another reason is that a very small change in base voltage can make a very big difference to the collector current.

Bad Circuits

Here are two variants of a very common circuit - which generally works **very badly**.

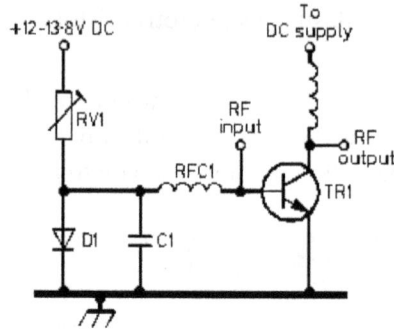

(a)

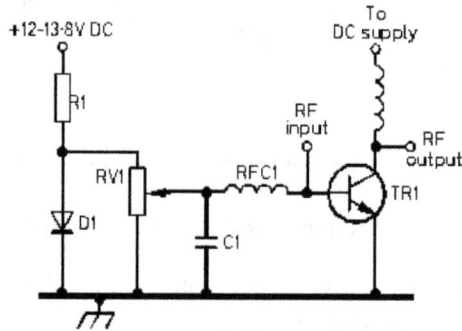

(b)

The voltage drop across the forward-biased diode D1 is approximately the same as the voltage drop across the transistor's base-emitter junction, and by adjusting RV1 to send the correct current through D1, the collector current in TR1 can be set to the desired value.

If D1 is in thermal contact with TR1, the voltage across D1 will drop as TR1 and D1 heat up together, maintaining a constant and temperature-stabilized current through TR1.

There are many problems with this simple circuit. The main one is that in order to maintain a constant voltage across D1, the permanent standing current through D1 must be several times higher than the maximum base current drawn by TR1 at the peak of RF drive. This requires an enormous standing current through RV1 and D1 - one ampere or even more. Most designers fail to provide this. Some commercial circuits even use a low-current signal diode such as a 1N914 for D1! Bias regulation totally fails when peaks of RF drive create a heavy demand for base current, so this circuit is a real splatter-generator.

Faithful Amplification

The process of raising the strength of a weak signal without any change in its general shape is known as faithful amplification. Conditions required for faithful amplification:

1. Proper zero signal collector current

2. Minimum proper base- emitter voltage at any instant

3. The Minimum proper collector-emitter voltage at ay instant

In short, the base-emitter junction shall remain properly forward biased and the base-collector junction shall remain properly reverse biased to achieve Faithful Amplicaiton

Base Resistor/Fixed Bias Circuit

The first biasing method, called BASE CURRENT BIAS or sometimes FIXED BIAS, was used in the picture below. As you recall, it consisted basically of a resistor (RB) connected between the collector supply voltage and the base. Unfortunately, this simple arrangement is quite thermally unstable. If the temperature of the transistor rises for any reason (due to a rise in ambient temperature or due to current flow through it), collector current will increase. This increase in current also causes the dc operating point, sometimes called the quiescent or static point, to move away from its desired position (level).

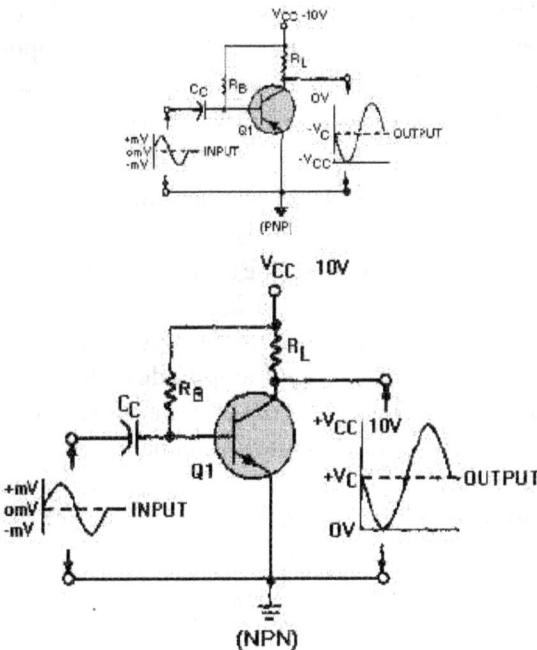

Biasing With Feedback Resistor

If some percentage of an amplifier's output signal is connected to the input, so that the amplifier amplifies part of its own output signal, we have what is known as *feedback*. Feedback comes in two varieties: *positive* (also called *regenerative*), and *negative* (also called *degenerative*). Positive feedback reinforces the direction of an amplifier's output voltage change, while negative feedback does just the opposite.

A familiar example of feedback happens in public-address ("PA") systems where someone holds the microphone too close to a speaker: a high-pitched "whine" or "howl" ensues, because the audio amplifier system is detecting and amplifying its own noise. Specifically, this is an example of *positive* or *regenerative* feedback, as any sound detected by the microphone is amplified and turned into a louder sound by the speaker, which is then detected by the microphone again, and so on... the result being a noise of steadily increasing volume until the system becomes "saturated" and cannot produce any more volume.

One might wonder what possible benefit feedback is to an amplifier circuit, given such an annoying example as PA system "howl." If we introduce positive, or regenerative, feedback into an amplifier circuit, it has the tendency of creating and sustaining oscillations, the frequency of which determined by the values of components handling the feedback signal from output to input.

Negative feedback, on the other hand, has a "dampening" effect on an amplifier: if the output signal happens to increase in magnitude, the feedback signal introduces a decreasing influence into the input of the amplifier, thus opposing the change in output signal. While positive feedback drives an amplifier circuit toward a point of instability (oscillations), negative feedback drives it in the opposite direction: toward a point of stability.

An amplifier circuit equipped with some amount of negative feedback is not only more stable, but it distorts the input waveform less and is generally capable of amplifying a wider range of frequencies. The tradeoff for these advantages (there just *has* to be a disadvantage to negative feedback, right?) is decreased gain. If a portion of an amplifier's output signal is "fed back" to the input to oppose any changes in the output, it will require a greater input signal amplitude to drive the amplifier's output to the same amplitude as before. This constitutes a decreased gain. However, the advantages of stability, lower distortion, and greater bandwidth are worth the tradeoff in reduced gain for many applications. Let's examine a simple amplifier circuit and see how we might introduce negative feedback into it, starting with the Figure 3.13.

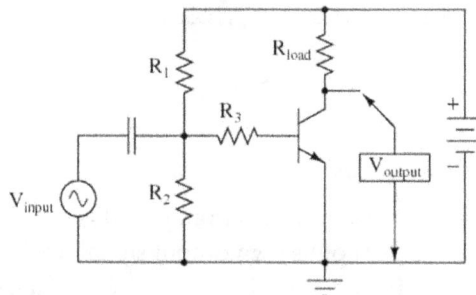

Figure 3.13: Amplifier Circuit.

Common-Emitter Amplifier without Feedback

The amplifier configuration shown here is a common-emitter, with a resistor bias network formed by R_1 and R_2. The capacitor couples V_{input} to the amplifier so that the signal source doesn't have a DC voltage imposed on it by the R_1/R_2 divider network. Resistor R_3 serves the purpose of controlling voltage gain. We could omit it for maximum voltage gain, but since base resistors like this are common in common-emitter amplifier circuits, we'll keep it in this schematic.

Like all common-emitter amplifiers, this one *inverts* the input signal as it is amplified. In other words, a positive-going input voltage causes the output voltage to decrease, or move toward the negative, and vice versa. The oscilloscope waveforms are shown in Figure 3.14.

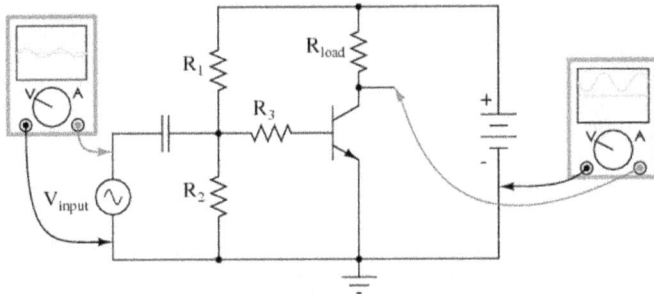

Figure 3.14: Oscilloscope Waveforms.

Because the output is an inverted, or mirror-image, reproduction of the input signal, any connection between the output (collector) wire and the input (base) wire of the transistor in Figure 3.15 will result in *negative* feedback.

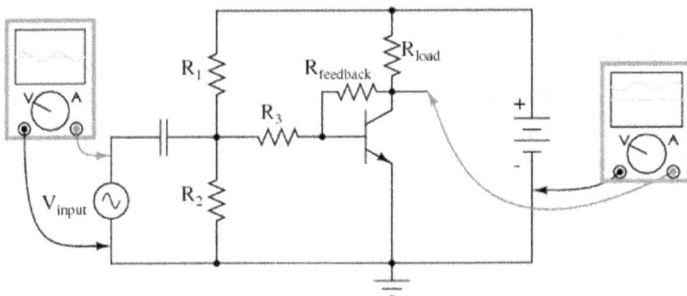

Figure 3.15: Common-Emitter Amplifier without Feedback.

The resistances of R_1, R_2, R_3, and $R_{feedback}$ function together as a signal-mixing network so that the voltage seen at the base of the transistor (with respect to ground) is a weighted average of the input voltage and the feedback voltage, resulting in signal of a reduced amplitude going into the transistor. So, the amplifier circuit in Figure above will have reduced voltage gain, but improved linearity (reduced distortion) and increased bandwidth.

Self Biasing

A better method of biasing is obtained by inserting the bias resistor directly between the base and collector, as shown in Figure 3.16. By tying the collector to the base in this manner, feedback voltage can be fed from the collector to the base to develop a forward bias. This arrangement is called SELF-BIAS.

Now, if an increase of temperature causes an increase in collector current, the collector voltage (VC) will fall because of the increase of voltage produced across the load resistor (RL). This drop in VC will be fed back to the base and will result in a decrease in the base current. The decrease in base current will oppose the original increase in collector current and tend to stabilize it. The exact opposite effect is produced when the collector current decreases.

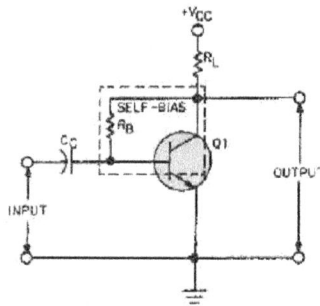

Figure 3.16: Self Biasing.

Self-bias has two small drawbacks: (1) It is only partially effective and, therefore, is only used where moderate changes in ambient temperature are expected; (2) it reduces amplification since the signal on the collector also affects the base voltage.

This is because the collector and base signals for this particular amplifier configuration are 180 degrees out of phase (opposite in polarity) and the part of the collector signal that is fed back to the base cancels some of the input signals. This process of returning a part of the output back to its input is known as DEGENERATION or NEGATIVE FEEDBACK. Sometimes degeneration is desired to prevent amplitude distortion (an output signal that fails to follow the input exactly) and self-bias may be used for this purpose.

Voltage or Potential Divider Biasing

This is the most widely used biasing scheme in general electronics. For a single stage amplifier, this circuit offers the best resilience against changes in temperature and device characteristics. The disadvantage is that a couple of extra resistors are required, but this is outweighed by the advantage of excellent stability. The biasing circuits is given in Figure 3.17.

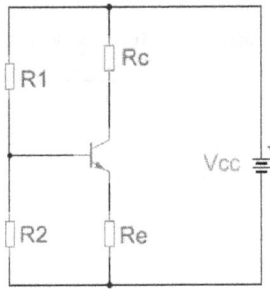

Figure 3.17: Biasing Circuit.

Here R1 and R2 form a potential divider, which will fix the base potential of the transistor. The current through this bias chain is usually set at 10 times greater than the base current required by the transistor. The base- emitter voltage drop of the transistor is approximated as 0.6 volt. There will also be a voltage drop across the emitter resistor, Re, this is generally set to about 10% of the supply voltage. The inclusion of this resistor also helps to stabilize the bias:

If the temperature increases, then extra collector current will flow. If Ic increases, then so will Ie as Ie = Ib + Ic. The extra current flow through Re increases the voltage drop across this resistor reducing the effective base- emitter voltage and therefore stabilizing the collector current. The equations follow:

Rc = Vc / Ic

Ie = Ib + Ic as Ic >> Ib then Ie ~ Ic

Ve = 10% * Vcc

Re = Ve / Ie

Vb = Ve + 0.6

R2 = Vb / 10 * Ib

R1 = Vcc-Vb / 10 * Ib

Battery Bias Stabilisation

Transistors are inevitable parts of Electronic circuits. The success of a circuit design lies in the selection of proper transistor type and calculation of voltage and the current flowing through it. A small variation in the voltage or current level in the transistor will affects the working of the whole circuit.

The Figure 3.18 explains how voltage and current are flowing through a bipolar transistor. The input voltage to the circuit is 12 volt DC. The base of T1 is connected to a potential divider R1-R2. If they have equal values, the half supply voltage will be available at the base of T1. Here the value of R1 is **3.2 Ohms**. If the value of R1 is three times greater than R2, then three- quarter of 12V drops by R1 and allow one quarter to pass through R2. Therefore the base voltage of T1 will be 12 / 4 = 3 V.

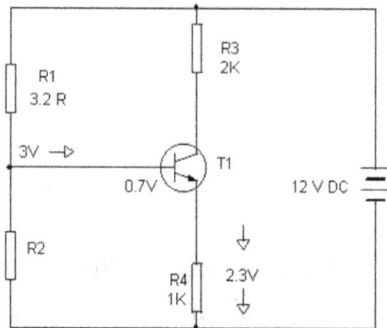

Figure 3.18: Flow of Voltage and Current Through a Bipolar Transistor.

Thus the voltage provided by R1 to the base of T1 is 3 volts. The emitter voltage of T1 will be 0.7 volts less than 3 volts since T1 drops 0.7 volts for its biasing. Thus the emitter voltage appears as 3 - 0.7 = 2.3 volts. If the value of the emitter resistor R4 is 1K, then if 2.3 volt passes through it, emitter current will be 2.3V/ 1 = 2.3 mA. Collector current also remains same. If the value of the load resistor R3 is 2K, two times higher than that of R4, then the voltage drop across it will be 2 × 2.3V = 4.6 volts, therefore the collector voltage of T1 remains as 12 – 4.6 = 7.4 volts.

Load Current

In the circuit shown in Figure 3.19 volt DC supply is provided. T1 is a general purpose NPN transistor like BC 548. A potential divider comprising R1 and R2 bias the base of T1. The minimum base voltage necessary for biasing T1 is 0.7 volts. The potential divider R1 - R2 drops 6 - 0.7 = 6.3 volts. If the load takes 4 volts, then the collector voltage will be 2 volts. **6 - 4 = 2 volts.**

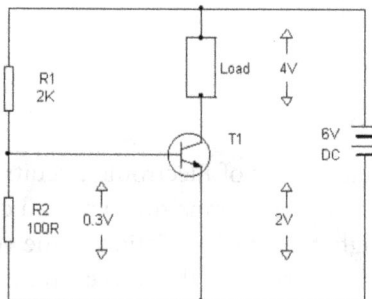

Figure 3.19: NPN Transistor.

Value of the collector current depends on the base voltage. When the base voltage increases, collector current also increases. This results in more volts in the load. In short, 0.1 volt increase in base voltage causes 1 Volt increases in the load.

Bias Compensation

The Minimum Statistics noise power spectral density (psd) estimation approach is based on tracking minima of a short term power spectral density (psd) estimate in frequency sub-bands. Since the short term minimum power is always smaller than (or in trivial cases equal to) the mean power, the minimum noise power estimate is a biased estimate of the mean power. For an accurate noise power estimate, this bias must be compensated.

Design of Biasing Circuit

A bipolar junction transistor (BJT) is very versatile. It can be used in many ways, as an amplifier, a switch or an oscillator and many other uses too. Before an input signal is applied its operating conditions need to be set. This is achieved with a suitable bias circuit, some of which I will describe. A bias circuit allows the operating conditions of a transistor to be defined so that it will operate over a predetermined range. This is normally achieved by applying a small fixed dc voltage to the input terminals of a transistor.

Bias design can take a mathematical approach or can be simplified using transistor characteristic curves. The characteristic curves predict the performance of a BJT. There are three curves, an input characteristic curve, a transfer characteristic curve and an output characteristic curve. Of these curves, the most useful for amplifier design is the output characteristics curve. The output characteristic curves for a BJT are a graph displaying the output voltages and currents for different input currents. The linear (straight) part of the curve needs is utilized for an amplifier or oscillator. For use as a switch, a transistor is biased at the extremities of the graph, these conditions are known as "cut-off" and "saturation".

For each transistor configuration, common emitter, common base, and emitter follower the output curves are slightly different. A typical output characteristic for a BJT in common emitter mode is shown in Figure 3.20.

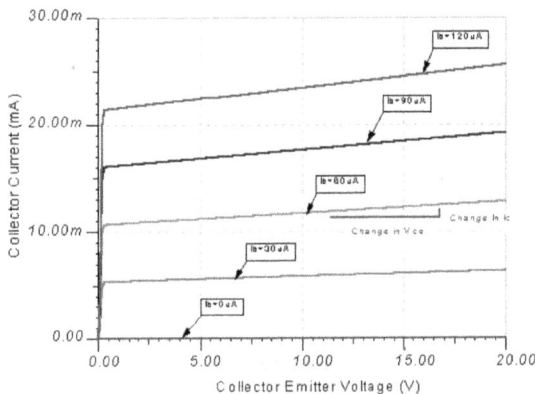

Figure 3.20: BJT in Common Emitter Mode.

After the initial bend, the curves approximate a straight line. The slope or gradient of each line represents the output impedance, for a particular input base current. So what has all this got to do with biasing? Take, for example the middle curve. The collector emitter voltage is displayed up to 20 volts. Let's assume that we have a single stage amplifier, working in common emitter mode, and the supply voltage is 10 volts. The output terminal is the collector, the input is the base, where do you set the bias conditions? The answer is anywhere on the flat part of the graph. However, imagine the bias is set so that the collector voltage is 2 volts. What happens if the output signal is 4 volts peak to peak? Depending on whether the transistor used is a PNP or NPN, then one half cycle will be amplified cleanly, the other cycle will approach the limits of the power supply and will "clip" as shown in Figure 3.21.

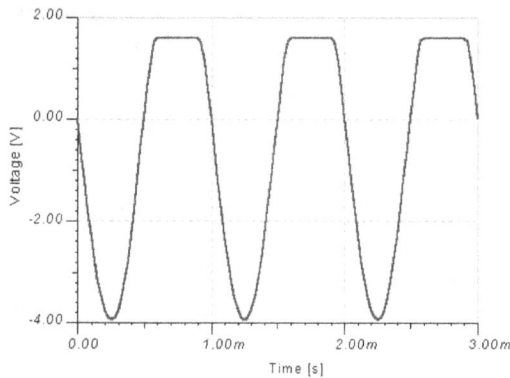

Figure 3.21: Transistor Waveform.

The Figure 3.21 shows a 4 volt peak to peak waveform with clipping on the positive half cycle. This is caused by setting the bias at a value other than half the supply voltage.

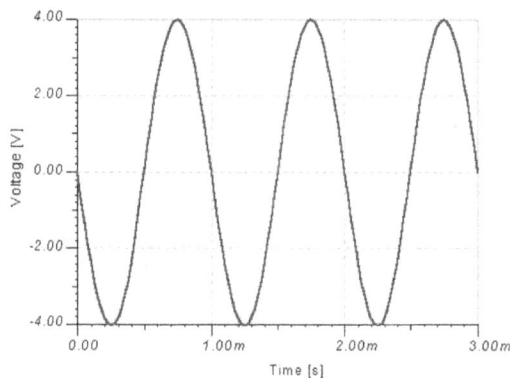

Figure 3.22

The Figure 3.22 shows the same amplifier, but here the bias is set so that collector voltage is half the value of the supply voltage. Hence, it is a good idea to set the bias for a single stage amplifier to half the supply voltage, as this allows maximum output voltage swing in both directions of an output waveform.

Before describing the bias circuits, it is worthwhile looking at a typical input characteristic curve for a small signal BJT. The following is the input characteristic for a transistor in common emitter mode; it is a plot of input base- emitter voltage versus base current. It is shown with both x and y- axis slightly zoomed (Figure 3.23).

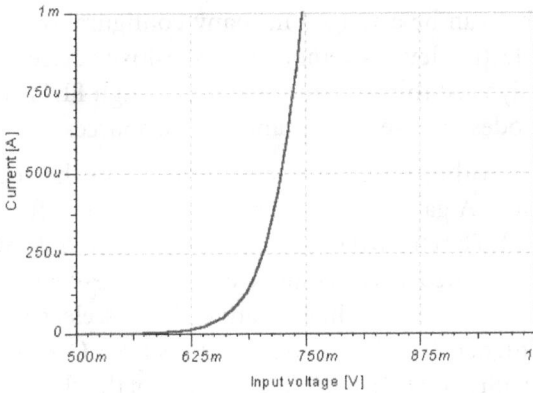

Figure 3.23: V-I Characteristics.

The base- emitter voltage, Vbe is found as either 0.6 V or 0.7 V. Both values are an approximation, and as can be seen from the above graph the value of Vbe varies between this range. For small signal work with base currents of 50uA or below a value for Vbe of 0.6 volts is a reasonable quote. For higher base currents, a Vbe of 0.7 V is a better approximation. In fact, in a large power transistor, the Vbe value can be even higher. The value of Vbe also varies widely with temperature change.

Circuit

The simplest bias circuit is shown in Figure 3.24. It consists only of a fixed bias resistor and load resistor. The BJT is operating in common emitter mode. The DC current gain or beta, h_{FE} is the ratio of dc collector current divided by dc base current. The BJT is a BC107A. The values of Rb and Rc can be determined by either mathematical approach or by using the output characteristic curves for the BC107A.

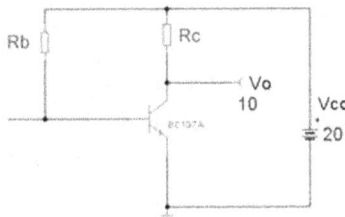

Figure 3.24: BJT is a BC107A

FET

Introduction

The field-effect transistor (FET) controls the current between two points but does so differently than the bipolar transistor. The FET operates by the effects of an electric field on the flow of electrons through a single type of semiconductor material. This is why the FET is sometimes called a unipolar transistor. Also, unlike bipolar semiconductors that can be arranged in many configurations to provide diodes, transistors, photoelectric devices. temperature sensitive devices and so on, the field effect is usually only used to make transistors, although FETs are also available as special-purpose diodes, for use as constant current sources.

Current moves within the FET in a channel, from the source connected to the drain connection. A gate terminal generates an electric field that controls the current (see Fig 8.25). The channel is made of either N-type or P-type semiconductor material; a FET is specified as either an N-channel or P-channel device. Majority carriers flow from source to drain. In N-channel devices, electrons flow so the drain potential must be higher than that of the Source (VDS > O)- In P-channel devices, the flow of holes requires that VDS < 0. The polarity of the electric field that controls current in the channel is determined by the majority carriers of the channel, ordinarily positive for P-channel FETs and negative for N-channel FETS.

Variations of FET technology are based on different ways of generating the electric field. In all of these, however, electrons at the gate are used only for their charge in order to create an electric field around the channel, and there is a minimal flow of electrons through the gate. This leads to a very high de input resistance in devices that use FETs for their input circuitry. There may be quite a bit of capacitance between the gate and the other FET terminals, however. The input impedance may be quite low at RF.

The current through an FET only has to pass through a single type of semiconductor material. There is very little resistance in the absence of an electric field (no bias voltage). The drain-source resistance (rDS ON) is between a few hundred ohms to less than an ohm. The output impedance of devices made with FETs is generally quite low. If a gate bias voltage is added to operate the transistor near cut off, the circuit output impedance may be much higher.

FET devices are constructed on a substrate of doped semiconductor material. The channel is formed within the substrate and has the opposite polarity (a P-channel FET has N-type substrate). Most FETS are constructed with silicon. In order to achieve a higher gain-bandwidth product, other materials have been used. Gallium Arsenide (GaAs) has electron mobility and drift velocities that are far higher than the standard doped silicon, Amplifiers designed with GaAs FET devices have much higher frequency response and lower noise factor at VHF and UHF than those made with standard FETS.

Construction of JEFT

All FETs have a gate, drain, and source terminals that correspond roughly to the base, collector, and emitter of BJTs. Most FETs also have a fourth terminal called the body, base, bulk, or substrate. This fourth terminal serves to bias the transistor into operation; it is rare to make non-trivial use of the body terminal in circuit designs, but its presence is important when setting up the physical layout of an integrated circuit. The size of the gate, length L in the diagram, is the distance between source and drain. The width is the extension of the transistor, in the diagram perpendicular to the cross- section. Typically the width is much larger than the length of the gate. A gate length of 1 μm limits the upper frequency to about 5 GHz, 0.2 μm to about 30 GHz.

The names of the terminals refer to their functions. The gate terminal may be thought of as controlling the opening and closing of a physical gate. This gate permits electrons to flow through or blocks their passage by creating or eliminating a channel between the source and drain. Electrons flow from the source terminal towards the drain terminal if influenced by an applied voltage. The body simply refers to the bulk of the semiconductor in which the gate, source and drain lie. Usually, the body terminal is connected to the highest or lowest voltage within the circuit, depending on type. The body terminal and the source terminal are sometimes connected together since the source is also sometimes connected to the highest or lowest voltage within the circuit, however, there are several uses of FETs which do not have such a configuration, such as transmission gates and cascode circuits (Figure 3.25).

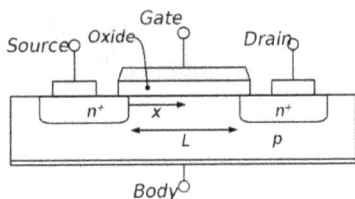

Figure 3.25: FET Cascode Circuits.

The FET controls the flow of electrons (or electron holes) from the source to drain by affecting the size and shape of a "conductive channel" created and influenced by voltage (or lack of voltage) applied across the gate and source terminals. This conductive channel is the "stream" through which electrons flow from source to drain.

In an n-channel depletion-mode device, a negative gate-to-source voltage causes a depletion region to expand in width and encroach on the channel from the sides, narrowing the channel. If the depletion region expands to completely close the channel, the resistance of the channel from source to drain becomes large, and the FET is effectively turned off like a switch. Likewise, a positive gate-to-source voltage increases the channel size and allows electrons to flow easily.

Conversely, in an n-channel enhancement-mode device, a positive gate-to-source voltage is necessary to create a conductive channel, since one does not exist naturally within the transistor. The positive voltage attracts free-floating electrons within the body towards the gate, forming a conductive channel. But first, enough electrons must be attracted near the gate to counter the dopant ions added to the body of the FET; this forms a region free of mobile carriers called a depletion region, and the phenomenon is referred to as the threshold voltage of the FET. Further gate-to-source voltage increase will attract even more electrons towards the gate which are able to create a conductive channel from source to drain; this process is called the inversion.

For either enhancement- or depletion-mode devices, at drain-to-source voltages much less than gate-to-source voltages, changing the gate voltage will alter the channel resistance, and drain current will be proportional to drain voltage (referenced to source voltage). In this mode the FET operates like a variable resistor and the FET is said to be operating in a linear mode or ohmic mode.

If the drain-to-source voltage is increased, this creates a significant asymmetrical change in the shape of the channel due to a gradient of voltage potential from source to drain. The shape of the inversion region becomes "pinched-off" near the drain end of the channel. If the drain-to-source voltage is increased further, the pinch-off point of the channel begins to move away from the drain towards the source. The FET is said to be in saturation mode; some authors refer to it as active mode, for a better analogy with bipolar transistor operating regions. The saturation mode, or the region between Ohmic and saturation, is used when amplification is needed. The in-between region is sometimes considered to be part of the ohmic or linear region, even where drain current is not approximately linear with drain voltage.

Even though the conductive channel formed by gate-to-source voltage no longer connects the source to drain during saturation mode, carriers are not blocked from flowing. Considering again an n-channel device, depletion regions exists in the p-type body, surrounding the conductive channel and drain and source regions. The electrons which comprise the channel are free to move out of the channel through the depletion region if attracted to the drain by drain-to-source voltage. The depletion region is free of carriers and has a resistance similar to silicon. Any increase of the drain-to-source voltage will increase the distance from drain to the pinch-off point, increasing resistance due to the depletion region proportionally to the applied drain-to-source voltage. This proportional change causes the drain-to-source current to remain relatively fixed independently of changes to the drain-to-source voltage and quite unlike the linear mode of operation. Thus in saturation mode, the FET behaves as a constant-current source rather than as a resistor and can be used most effectively as a voltage amplifier. In this case, the gate-to-source voltage determines the level of constant current through the channel.

Biasing of FET

Biasing in electronics is the method of establishing predetermined voltages and/ or currents at various points of an electronic circuit to set an appropriate operating point. The operating point of a device, also known as bias point, quiescent point, or simply Q-point, is the point on the output characteristics that shows the direct current (DC), collector-emitter voltage (V_{CE}), and the collector current (I_C) with no input signal applied. The term is normally used in connection with devices such as transistors.

For bipolar junction transistors, the bias point is chosen to keep the transistor operating in the active mode, using a variety of circuit techniques, establishing the Q-point DC voltage and current. A small signal is then applied on top of the Q-point bias voltage, thereby either modulating or switching the current, depending on the purpose of the circuit. The quiescent point of operation is typically near the middle of DC load line. The process of obtaining certain DC collector current at a certain DC collector voltage by setting up operating point is called biasing.

After establishing the operating point, when the input signal is applied, the output signal should not move the transistor either to saturation or to cut-off. However, this unwanted shift still might occur, due to the following reasons:

1. Parameters of transistors depend on junction temperature. As junction temperature increases, leakage current due to minority charge carriers (I_{CBO}) increases. As I_{CBO} increases, I_{CEO} also increases, causing an increase in collector current I_C. This produces heat at the collector junction. This process repeats, and, finally, Q-point may shift into the saturation region. Sometimes, the excess heat produced at the junction may even burn the transistor. This is known as thermal runaway.

2. When a transistor is replaced by another of the same type, the Q-point may shift, due to changes in parameters of the transistor, such as *current gain* (β) which varies slightly for each unique transistor.

To avoid a shift of Q-point, bias-stabilization is necessary. Various biasing circuits can be used for this purpose.

Working Principle of N- FET

The function of Field Effect Transistors is similar to bipolar transistors but there are a few differences. They have 3 terminals as shown below. Two general types of FETs are the 'N' channel and the 'P' channel MOSFETs. Here we will only discuss the N channel. Actually, in this section, we'll only be discussing the most commonly used enhancement mode N channel MOSFET (Metal Oxide Semiconductor Field Effect Transistor). Its schematic symbol the arrows show how the LEGS of the actual transistor correspond to the schematic symbol (Figure 3.26).

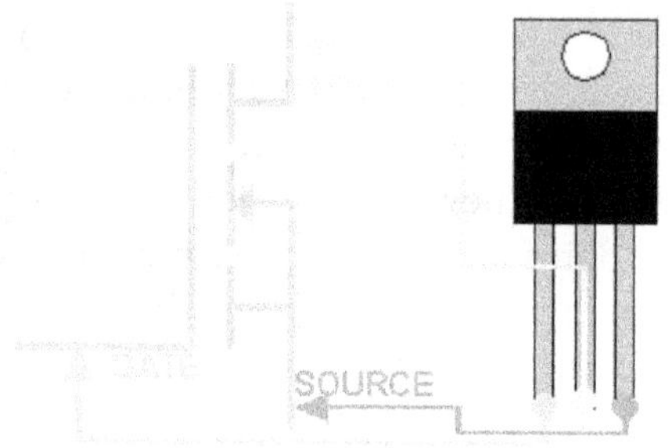

Figure 3.26: Metal Oxide Semiconductor Field Effect Transistor.

Current Control

The control terminal is called the gate. Remember that the base terminal of a bipolar transistor passes a small amount of current. The gate on the FET passes virtually no current when driven with D.C. When driving the gate with high-frequency pulsed D.C. or A.C. there may be a small amount of current flow. The transistor's "turn on" (a.k.a. threshold) voltage varies from one FET to another but is approximately 3.3 volts with respect to the source.

When FETs are used in the audio output section of an amplifier, the Vgs (voltage from gate to source) is rarely higher than 3.5 volts. When FETs are used in switching power supplies, the Vgs is usually much higher (10 to 15 volts). When the gate voltage is above approximately 5 volts, it becomes more efficient (which means less voltage drop across the FET and therefore less power dissipation).

MOSFETS are commonly used because they are easier to drive in high current applications (such as the switching power supplies found in car audio amplifiers). If a bipolar transistor is used, a fraction of the collector/emitter current must flow through the base junction. In high current situations where there is significant collector/emitter current, the base current may be significant. FETs can be driven by very little current (compared to the bipolar transistors). The only current that flows from the drive circuit is the current that flows due to the capacitance.

As you already know, when DC is applied to a capacitor, there is an initial surge then the current flow stops. When the gate of a FET is driven with a high- frequency signal, the drive circuit essentially sees only a small value capacitor. For low to intermediate frequencies, the drive circuit has to deliver little current. At very high frequencies or when many FETs are being driven, the drive circuit must be able to deliver more current.

Static Characteristic of FET

There are two types of static characteristics *viz.*

1. The output or drain characteristic.
2. Transfer characteristic.

Output or Drain Characteristic

The curve is drawn between drain current I_p and drain-source voltage V_{DS} with gate-to- source voltage V_{GS} as the parameter is called the *drain* or *output characteristic*. This characteristic is analogous to collector characteristic of a BJT (Figure 3.27):

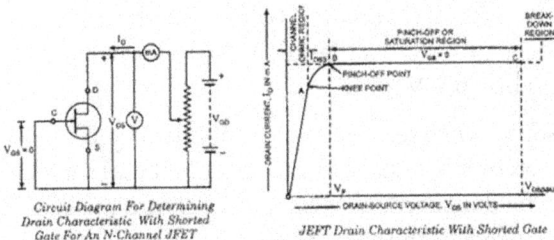

Circuit Diagram For Determining Drain Characteristic With Shorted Gate For An N-Channel JFET

JEFT Drain Characteristic With Shorted Gate

Figure 3.27: Drain *or the* Output Characteristic.

(a) Drain Characteristic with Shorted-Gate

The circuit diagram for determining the drain characteristic with shorted-gate for an N-channel JFET is given in the figure and the drain characteristic with shorted-gate is shown in another figure.

Initially, when drain-source voltage V_{ns} is zero, there is no attracting potential at the drain, so no current flows in spite of the fact that the channel is fully open. This gives drain current $Ip = 0$. For small applied voltage V_{na}, the N-type bar acts as a simple semiconductor resistor, and the drain current increases linearly withthe increase in Vds, up to the knee point. This region, (to the left of the knee point) of the curve is called the channel Ohmic *region* because in this region the FET behaves like an ordinary resistor.

With the increase in drain current I_D, the ohmic voltage drop between the source and channel region reverse-biases the gate junction. The reverse-biasing of the gate junction is not uniform throughout. The reverse bias is more at the drain end than that at the source end of the channel, so with the increase in Vds, the conducting portion of the channel begins to constrict more at the drain end. Eventually, a voltage Vds is reached at which the channel is pinched off. The drain current I_D no longer increases with the increase in Vds. It approaches a constant saturation value. The value of voltage V_{DS} at which the channel is pinched off (*i.e.* all the free charges from the channel get removed), is called the *pinch-off voltage* V_p. The pinch-off voltage Vp, not too sharply defined on the curve, where the drain current I_D begins to level off and attains a constant value. From point A (knee point) to the point B (pinch-off point) the drain current I_D increases with the increase In voltage V_{ds} following a reverse square law. The region of the characteristic in which drain current I_D remains fairly constant is called *the pinch-off region*. It is also sometimes called the *saturation region* or *amplifier region*. In this region, the JFET operates as a *constant current device since* drain current (or output current) remains almost constant. It is the normal operating region of the JFET when used as an amplifier. The drain current in the pinch-off region with $V_{GS} = 0$ is referred to the *drain-source saturation current, Idss*).

It is to be noted that in the pinch-off (or saturation) region the channel resistance increases in proportion to increase in V_{DS} and so keeps the drain current almost constant and the reverse bias required by the gate-channel junction is supplied entirely by the voltage drop across the channel resistance due to flow of I_{Dsg} and not by the external bias because $V_{GS} = 0$.

If the drain-source voltage, Vds is continuously increased, a stage comes when the gate-channel junction breaks down. At this point current increases very rapidly and the JFET may be destroyed. This happens because the charge carriers making up the saturation current at the gate channel junction accelerate to a high velocity and produce an *avalanche effect* (Figure 3.28).

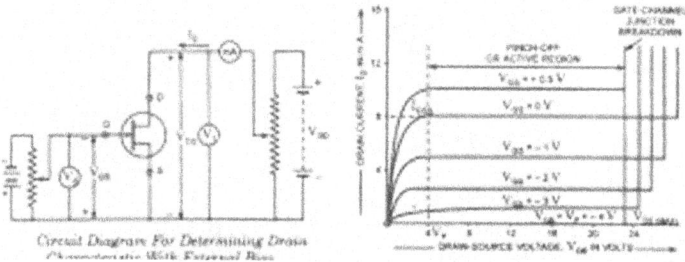

Figure 3.28: JFET-Drain Characteristics with External Bias.

Transfer Characteristic of JFET

The transfer characteristic for a JFET can be determined experimentally, keeping drain-source voltage, *VDS* constant and determining drain current, I_D for various values of gate-source voltage, V_{GS}. The circuit diagram is shown in fig. The curve is plotted between gate-source voltage, V_{GS} and drain current, I_D, as illustrated in fig. It is similar to the trans-conductance characteristic of a vacuum tube or a transistor. It is observed that

(i) Drain current decreases with the increase in negative gate-source bias

(ii) Drain current, $I_D = I_{DSS}$ when $V_{GS} = 0$

(iii) Drain current, $I_D = 0$ when $V_{GS} = V_D$ The transfer characteristic follows equation

The transfer characteristic can also be derived from the drain characteristic by noting values of drain current, I_D corresponding to various values of gate-source voltage, V_{GS} for a constant drain-source voltage and plotting them.

It may be noted that a P-channel JFET operates in the same way and have the similar characteristics as an N-channel JFET except that channel carriers are holes instead of electrons and the polarities of V_{GS} and V_{DS} are reversed.

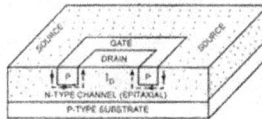

Single-Ended-Geometry JFET

Merits/Demerits

Junction field effect transistors combine several merits of both conventional (or bipolar) transistors and vacuum tubes. Some of these are enumerated below:

1. Its operation depends upon the flow of majority carriers only; it is, therefore, a unipolar (one type of carrier) device. On the other hand in an ordinary transistor, both majority and minority carriers take part in conduction and, therefore, the ordinary transistor is sometimes called the bipolar transistor. The vacuum tube is another example of a unipolar device. '

2. It is simpler to fabricate, smaller in size, rugged in construction and has a longer life and higher efficiency. Simpler to fabricate in IC form and space requirement is also lesser.

3. It has a high input impedance (of the order of 100 M Q), because its input circuit (gate to source) is reverse biased, and so permits a high degree of isolation between the input and the output circuits. However, the input circuit of an ordinary transistor is forward biased and, therefore, an ordinary transistor has low input impedance.

4. It carries very small current because of the reverse biased gate and, therefore, it operates just like a vacuum tube where control grid (corresponding to the gate in JFET) carries extremely small current and input voltage controls the output current. This is the reason that JFET is essentially a voltage driven device.

5. An ordinary transistor uses a current into its base for controlling a large current between collector and emitter whereas in a JFET voltage on the gate (base) terminal is used for controlling the drain current (current between drain and source). Thus an ordinary transistor gain is characterized by current gain whereas the JFET gain is characterized as the trans-conductance (the ratio of drain current and gate-source voltage).

6. JFET has no junction like an ordinary transistor and the conduction is through bulk material current carriers (N-type or P-type semiconductor material) that do not cross junctions. Hence the inherent noise of tubes (owing to high- temperature operation) and that of ordinary transistors (owing to junction transitions) is not present in JFET.

7. It is relatively immune to radiation.

8. It has a negative temperature coefficient of resistance and, therefore, has better thermal stability.

9. It has high power gain and, therefore, the necessity of employing driver stages is eliminated.

10. It exhibits no offset voltage at zero drain current and, therefore, makes an excellent signal chopper.

11. It has square law characteristics and, therefore, it is very useful in the tuners of radio and TV receivers.

12. It has got high- frequency response.

The Main Drawback of JFET

1. Its relative small gain-bandwidth product in comparison with that of a conventional transistor.

2. Greater susceptibility to damage in its handling.

3. JFET has low voltage gains because of small trans-conductance.

4. Costlier when compared to BJT's.

FET Parameters

Parameter	Definition
F1	Drain to source resistance $R_{ds}\,(\Omega)$
F2	Trans-conductance $G_m\,(1/\Omega)$
F3	Gate to source capacitance $C_{gs}\,(Farads)$
F4	Gate to drain capacitance $C_{gd}\,(Farads)$
F5	Drain to source capacitance $C_{ds}\,(Farads)$
F6	Drain bias current $I_d\,(Amps)$
F7	DC gate current $I_g\,(Amps)$
F8	Drain shot noise 1/f corner $(Hertz)$
F9	Gate shot noise 1/f corner $(Hertz)$

The noise sources are determined from the Gm, Id, and Ig parameters. See Appendix B for more information on FET noise sources.

FET Applications

Low Noise Amplifier

Noise is an undesirable disturbance super-imposed on a useful signal. Noise interferes with the information contained in the signal; the greater the noise, the less the information. For instance, the noise in radio-receivers develops crackling and hissing which sometimes completely masks the voice or music. Similarly, the noise in TV receivers produces small white or black spots on the picture; a severe noise may wipe out the picture. Noise is independent of the signal strength because it exists even when the signal is off.

Every electronic device produces a certain amount of noise but FET is a device which causes very little noise. This is especially important near the front-end of the receivers and other electronic equipment because the subsequent stages amplify front-end noise along with the signal. If FET is used at the front-end, we get less amplified noise (disturbance) at the final output.

Buffer Amplifier

A buffer amplifier is a stage of amplification that isolates the preceding stage from the following stage. Source follower (common drain) is used as a buffer amplifier. Because of the high input impedance, and low output impedance a FET acts as an excellent buffer amplifier, as shown in the Figure 3.29. Owing to high input impedance almost all the output voltage of the preceding stage appears at the input of the buffer amplifier and owing to low output impedance all the output voltage from the buffer amplifier reaches the input of the following stage, even there may be a small load resistance.

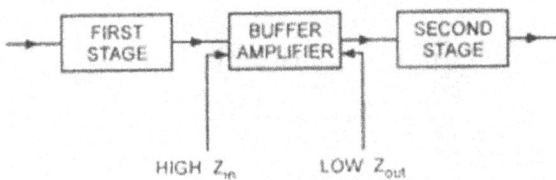

Figure 3.29: Buffer Amplifier.

Cascode Amplifier

Circuit diagram for a cascode amplifier using FET is shown in Figure 3.30. A common source amplifier drives a common gate amplifier in it.

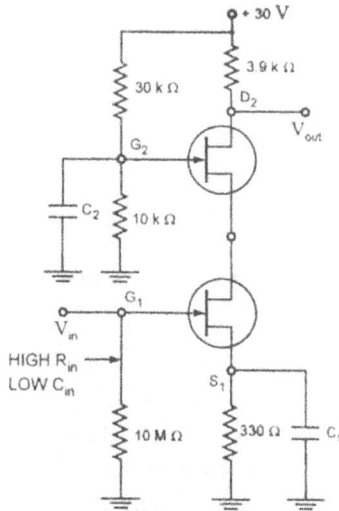

Figure 3.30: Cascode Amplifier.

The cascode amplifier has the same voltage gain as a common source (CS) amplifier. The main advantage of cascode connection is its low input capacitance which is considerably less than the input capacitance of a CS amplifier. It has high input resistance which is also a desirable feature.

Analog Switch

FET as an analog switch is shown in Figure 3.31. When no gate voltage is applied to the FET *i.e.* $V_{GS} = 0$, FET becomes saturated and it behaves like a small resistance usually of the value of less than 100 ohm and, therefore, output voltage becomes equal to

$$V_{OUT} = \{R_{DS}/ (R_D + R_{DS (ON)})\}^* V_{in}$$

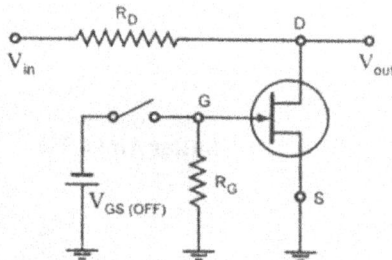

Figure 3.31: JFET as Analog Switch.

Since R_D is very large in comparison to $R_{DS\ ON}$), so V_{out} can be taken equal to zero.

When a negative voltage equal to $V_{GS(OFF)}$ is applied to the gate, the FET operates in the cut-off region and it acts like a very high resistance usually of some mega ohms. Hence output voltage becomes nearly equal to the input voltage.

Chopper

A direct-coupled amplifier can be built by leaving out the coupling and bypass capacitors and connecting the output of each stage directly to the input of next stage. Thus direct current is coupled, as well as alternating current. The major drawback of this method is the occurrence of drift, a slow shift in the final output voltage produced by supply transistor, and temperature variations. The drift problem can be overcome by employing chopper amplifier as illustrated in the Figure 3.32.

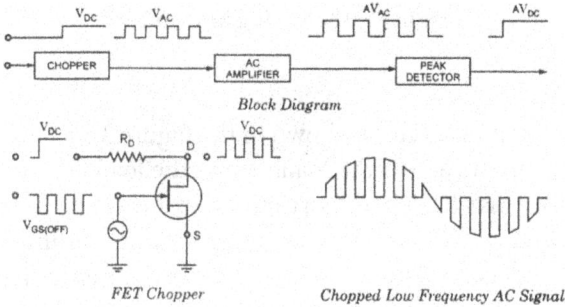

Figure 3.32: Chopper Amplifier Circuit.

Here input dc voltage is chopped by a switching circuit. The output of chopper is a square wave ac signal having peak value equal to that of input dc voltage, V_{DC}. This ac signal can be amplified by a conventional ac amplifier without any problem of drift. Amplified output can then be 'peak detected' to recover the amplified dc signal.

The Square wave is applied to the gate of a FET analog switch to make it operate like a chopper, as illustrated in other figures. The gate square wave is negative-going swing from 0 V to at least V_{GS} (off)-This alternately saturates and cuts-off the JFET. This output voltage is a square wave varying from $+V_{DC}$ to zero volts alternately.

If the input signal is a low-frequency ac signal, it gets chopped into the ac waveform as shown in the figure. This chopped signal can now be amplified by an ac amplifier that is drift free. The amplified signal can then be peak-detected to recover the original input low- frequency ac signal. Thus both dc and low- frequency ac signals can be amplified by using a chopper amplifier.

Multiplexer

An *analog multiplexer*, a circuit that steers one of the input signals to the output line, is shown in the Figure 3.33. In this circuit, each JFET acts as a single-pole-single-throw switch. When the control signals(Vv V_2 and V_3) are more negative than $V_{GS(OFF)}$ all input signals are blocked. By making any control voltage equal to zero, one of the inputs can be transmitted to the output. For instance, when V_x is zero, the signal obtained at the output will be sinusoidal. Similarly, when V_2 is zero, the signal obtained at the output will be triangular and when V_3 is zero, the output signal will be square-wave one. Normally, only one of the control signals is zero.

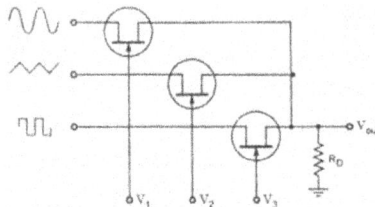

Figure 3.33: Analog Multiplier.

Current Limiter

JFET current limiting circuit is shown in the Figure 3.34. Almost all the supply voltage therefore, appears across the load. When the load current tries to increase to an excessive level (may be due to short-circuit or any other reason), the excessive load current forces the JFET into the active region, where it limits the current to 8 mA. The JFET now acts as a current source and prevents excessive load current. A manufacturer can tie the gate to the source and package the JFET as a two terminal device. This is how *constant-current diodes* are made. Such diodes are also called current-regulator diodes.

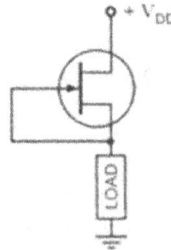

Figure 3.34: JFET Current Limiter Circuit.

Phase Shift Oscillators

JFET can incorporate the amplifying action as well as feedback action. It, therefore, acts well as a phase shift oscillator. The high input impedance of FET is especially very valuable in phase-shift oscillators in order to minimize the loading effect. A typical phase shift oscillator employing N-channel JFET is shown in the Figure 3.35.

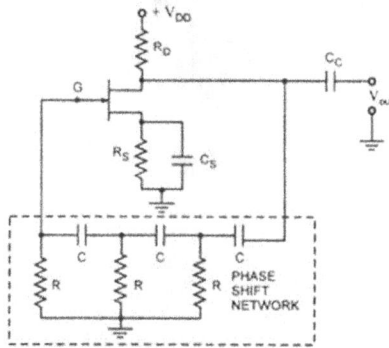

Figure 3.35: Phase Shift Oscillator.

Small Signal BJT Amplifier

Introduction

The current flowing between emitter and collector of a transistor is much greater than that flowing between base and emitter. So a small base current is controlling the much larger emitter- collector current. The ratio of the two currents, I_{CE} / I_{BE} is constant, provided that the collector emitter voltage V_{CE} is constant. Therefore, if the base current rises, so does collector current.

This ratio is the CURRENT GAIN of the transistor and is given the symbol h_{fe}. A fairly low gain transistor might have a current gain of 20 to 50, while a high gain type may have a gain of 300 to 800 or more. The spread of values of h_{fe} for any given transistor is quite large, even in transistors of the same type and batch.

Characteristic

The graph of I_{CE} / I_{BE} shown (right) is called the TRANSFER CHARACTERISTIC and the slope of the graph shows the h_{fe} for that transistor. Characteristic curves (graphs) can be drawn to show other parameters of a transistor, and are used both to detail the performance of a particular device and as an aid to the design of amplifiers (Figure 3.36).

The INPUT CHARACTERISTIC is a graph of base emitter current I_{BE} against base- emitter voltage V_{BE} (I_{BE}/V_{BE}) shows the input CONDUCTANCE of the transistor (Figure 3.37). As conductance ,I / V is the reciprocal of RESISTANCE (V / I) this curve can be used to determine the input resistance of the transistor. The steepness of this particular curve when the V_{BE} is above 1 volt shows that the input conductance is very high, and there is a large increase in current for a very small increase in V_{BE}. Therefore the input RESISTANCE must be low. Around 0.6 to 0.7 volts the graph curves shows that the input resistance of a transistor varies, according to the amount of base current flowing, and below about 0.5 volts base current ceases.

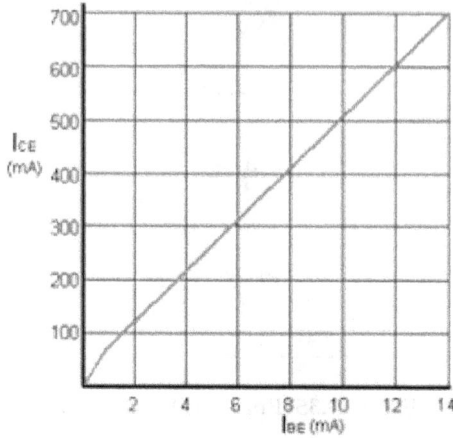

Figure 3.36: Transfer Characteristic.

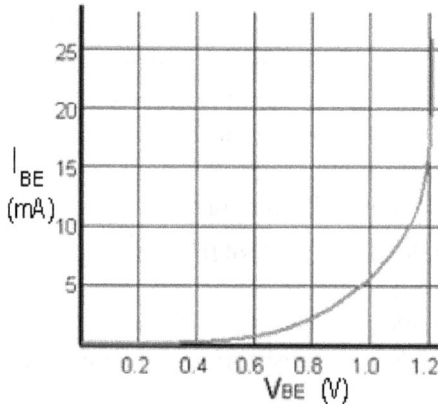

Figure 3.37: Input Characteristic.

Figure 3.38 shows the OUTPUT CHARACTERISTIC whose slope gives the value of output conductance (and by implication output resistance). The near horizontal parts of the graph lines show that a change in collector emitter voltage V_{CE} has almost no effect on collector current in this region, just the effect to be expected if the transistor output had a large value resistor in series with it. Therefore the graph shows that the output resistance of the transistor is high.

The above characteristic graphs show that for a silicon transistor with an input applied between base and emitter, and the output taken between collector and emitter (a method of connection called common emitter mode) one would expect;

☆ Low input resistance (from the input characteristic).

☆ Fairly high gain (from the transfer characteristic).

☆ High output resistance (from the output characteristic).

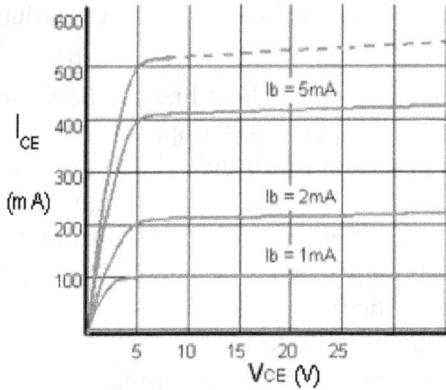

Figure 3.38: Output Characteristic.

Figure 3.39 shows, the MUTUAL CHARACTERISTIC shows a graph of MUTUAL CONDUCTANCE I_c/V_{BE} and illustrates the change in collector current which takes place for a given change in base- emitter voltage, (*i.e.* input signal voltage). This graph is for a typical silicon power transistor. Notice the large collector currents possible, and the nearly linear relationship between the input voltage and output current.

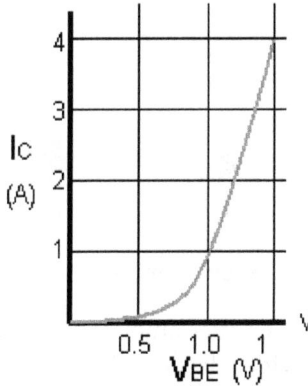

Figure 3.39: Mutual Characteristic.

Analysis of Transistor Amplifier

In the Bipolar Transistor tutorial, we saw that the most common circuit configuration for an NPN transistor is that of the Common Emitter Amplifier and that a family of curves known commonly as the Output Characteristics Curves, relates the Collector current (Ic), to the output or Collector voltage (Vce), for different values of Base current (Ib). All types of transistor amplifiers operate using AC signal inputs which alternate between a positive value and a negative value so some way of "presetting" the amplifier circuit to operate between these two maximum or peak values is required. This is achieved using a process known as Biasing. Biasing is very important in amplifier design as it establishes the correct operating point of the transistor amplifier ready to receive signals, thereby reducing any distortion to the output signal.

We also saw that a static or DC load line can be drawn onto these output characteristics curves to show all the possible operating points of the transistor from fully "ON" to fully "OFF", and to which the quiescent operating point or **Q-point** of the amplifier can be found. The aim of any small signal amplifier is to amplify all of the input signals with the minimum amount of distortion possible to the output signal, in other words, the output signal must be an exact reproduction of the input signal but only bigger (amplified). To obtain low distortion when used as an amplifier the operating quiescent point needs to be correctly selected. This is, in fact the DC operating point of the amplifier and its position may be established at any point along the load line by a suitable biasing arrangement. The best possible position for this Q-point is as close to the center position of the load line as reasonably possible, thereby producing a Class A type amplifier operation, *i.e.* Vce = 1/2Vcc. Consider the Common Emitter Amplifier circuit shown in Figure 3.40.

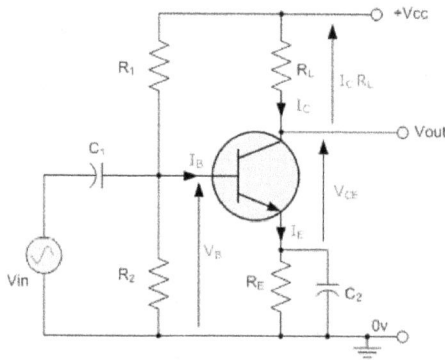

Figure 3.40: Common Emitter Amplifier Circuit.

The single stage common emitter amplifier circuit shown above uses what is commonly called "Voltage Divider Biasing". This type of biasing arrangement uses two resistors as a potential divider network and is commonly used in the design of bipolar transistor amplifier circuits.

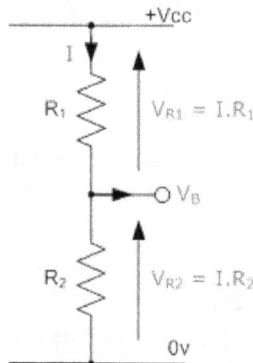

This method of biasing the transistor greatly reduces the effects of varying Beta, (β) by holding the Base bias at a constant steady voltage level allowing for best stability. The quiescent Base voltage (Vb) is determined by the potential divider network formed by the two resistors, R1, R2 and the power supply voltage Vcc as shown with the current flowing through both resistors. Then the total resistance RT will be equal toR1 + R2 giving the current as i = Vcc/RT. The voltage level generated at the junction of resistors R1 and R2 holds the Base voltage (Vb) constant at a value below the supply voltage. Then the potential divider network used in the common emitter amplifier circuit divides the input signal in proportion to the resistance. This bias reference voltage can be easily calculated using the simple voltage divider formula below:

$$V_B = \frac{V_{CC}\,R_2}{R_1 + R_2}$$

The same supply voltage, (Vcc) also determines the maximum Collector current, Ic when the transistor is switched fully "ON" (saturation), Vce = 0. The Base current Ib for the transistor is found from the Collector current, Ic and the DC current gain Beta, β of the transistor.

$$\beta = \frac{\Delta I_C}{\Delta I_B}$$

Beta is sometimes referred to as h_{FE} which is the transistors forward current gain in the common emitter configuration. Beta has no units as it is a fixed ratio of the two currents, Ic and Ib so a small change in the Base current will cause a large change in the Collector current. Transistors of the same type and part number will have large variations in their Beta value for, example, the BC107 NPN Bipolar transistor has a DC current gain Beta value of between 110 and 450 (datasheet value) this is because Beta is a characteristic of their construction and not their operation.

As the Base/Emitter junction is forward-biased, the Emitter voltage, Ve will be one junction voltage drop different to the Base voltage. If the voltage across the Emitter resistor is known then the Emitter current,Ie can be easily calculated using **Ohm's Law**. The Collector current, Ic can be approximated, since it is almost the same value as the Emitter current.

Data and Number Representation

Numerical Presentation

In science, technology, business, and, in fact, most other fields of endeavor, we are constantly dealing with quantities. Quantities are measured, monitored, recorded, manipulated arithmetically, observed, or in some other way utilized in most physical systems. It is important when dealing with various quantities that we be able to represent their values efficiently and accurately. There are basically **two** ways of representing the numerical value of quantities: *analog* and *digital*.

Analog Representation

In analog representation, a quantity is represented by a voltage, current, or meter movement that is proportional to the value of that quantity. Analog quantities such as those cited above have an important characteristic: they can vary over a continuous range of values (Figure 4.1).

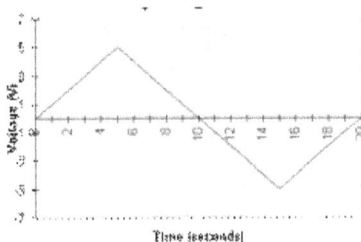

Figure 4.1: Graph of *Analog Voltage vs Time.*

Digital Representation

In digital representation, the quantities are represented not by proportional quantities but by symbols called digits. As an example, consider the digital watch, which provides the time of day in the form of decimal digits which represent hours and minutes (and sometimes seconds). As we know, the time of day changes continuously, but the digital watch reading does not change continuously; rather, it changes in steps of one per minute (or per second). In other words, this digital representation of the time of day changes in discrete steps, as compared with the representation of time provided by an analog watch, where the dial reading changes continuously (Figure 4.2).

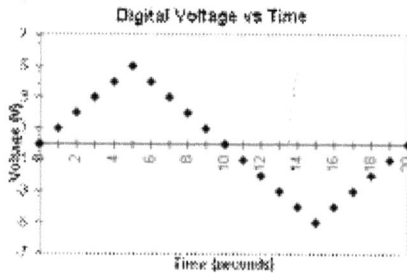

Figure 4.2: Graph of *Digital Voltage vs Time*.

The major difference between analog and digital quantities, then, can be simply stated as follows:

Analog = continuous

Digital = discrete (step by step)

Advantages and Limitations of Digital Techniques

Advantages

1. Easier to design. Exact values of voltage or current are not important, only the range (HIGH or LOW) in which they fall.
2. Information storage is easy.
3. Accuracy and precision are greater.
4. The operation can be programmed. Analog systems can also be programmed, but the variety and complexity of the available operations are severely limited.
5. Digital circuits are less affected by noise. As long as the noise is not large enough to prevent us from distinguishing a HIGH from a LOW.
6. More digital circuitry can be fabricated on IC chips.

Limitations

There is really only one major drawback when using digital techniques:

The real world is mainly analog.

Most physical quantities are analog in nature, and it is these quantities that are often the inputs and outputs that are being monitored, operated on, and controlled by a system.

To take advantage of digital techniques when dealing with analog inputs and outputs, three steps must be followed:

1. Convert the real-world analog inputs to digital form (ADC).
2. Process (operate on) the digital information.
3. Convert the digital outputs back to real-world analog form (DAC).

The following diagram shows a temperature control system that requires analog/digital conversions in order to allow the use of digital processing techniques.

Digital Number System

Many number systems are in use in digital technology. The most common is the decimal, binary, octal, and hexadecimal systems. The decimal system is clearly the most familiar to us because it is a tool that we use every day. Examining some of its characteristics will help us to better understand the other systems.

Decimal System

Decimal System The decimal system is composed of 10 numerals or symbols. These 10 symbols are 0, 1, 2, 3, 4, 5, 6, 7, 8, 9; using these symbols as digits of a number, we can express any quantity. The decimal system, also called the base-10 system because it has 10 digits.

10^3		10^2	10^1	10^0		10^{-1}	10^{-2}	10^{-3}
=1000		=100	=10	=1	.	=0.1	=0.01	=0.001
Most Significant Digit					Decimal point			Least Significant Digit

Binary System

In the binary system, there are only two symbols or possible digit values, 0 and 1. This base-2 system can be used to represent any quantity that can be represented in decimal or other number systems.

2^3		2^2	2^1	2^0		2^{-1}	2^{-2}	2^{-3}
=8		=4	=2	=1	.	=1/2	=1/4	=1/8
Most Significant Bit					Binary point			Least Significant Bit

Binary Counting

Table 4.1: Binary Counting Sequence

$2^3 = 8$	$2^2 = 4$	$2^1 = 2$	$2^0 = 1$	Decimal Equivalent
0	0	0	0	0
0	0	0	1	1
0	0	1	0	2
0	0	1	1	3
0	1	0	0	4
0	1	0	1	5
0	1	1	0	6
0	1	1	1	7
1	0	0	0	8
1	0	0	1	9
1	0	1	0	10
1	0	1	1	11
1	1	0	0	12
1	1	0	1	13
1	1	1	0	14
1	1	1	1	15

The binary counting sequence is depicted in Figure 4.3.

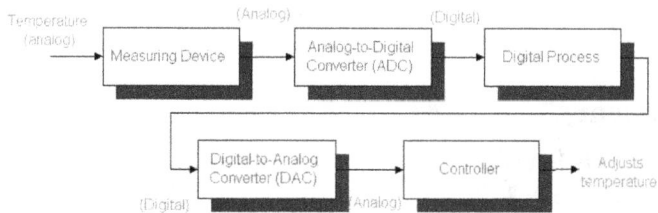

Figure 4.3: Binary Counting Sequence.

Representing Binary Quantities

In digital systems, the information that is being processed is usually presented in the binary form. Binary quantities can be represented by any device that has only two operating states or possible conditions, *e.g.* a switch has only open or closed. We arbitrarily (as we define them) let an open switch represent binary 0 and a closed switch represent binary 1. Thus we can represent any binary number by using series of switches.

Typical Voltage Assignment

Binary 1: Any voltage between 2V to 5V

Binary 0: Any voltage between 0V to 0.8V

Not used: Voltage between 0.8V to 2V, this may cause an error in a digital circuit.

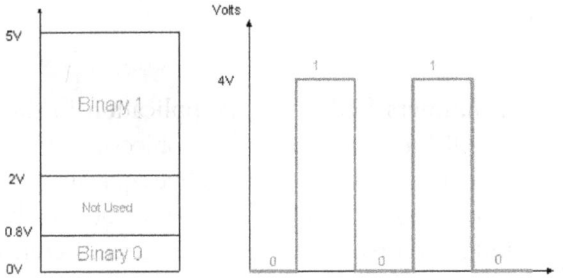

We can see another significant difference between digital and analog systems. In digital systems, the exact value of a voltage is *not* important; e.g., a voltage of 3.6V means the same as a voltage of 4.3V. In analog systems, the exact value of a voltage *is* important.

Number System and Codes

Numbers and Symbols

The expression of numerical quantities is something we tend to take for granted. This is both a good and a bad thing in the study of electronics. It is good, in that we're accustomed to the use and manipulation of numbers for the many calculations used in analyzing electronic circuits. On the other hand, the particular system of notation we've been taught from grade school onward is *not* the system used internally in modern electronic computing devices, and learning any different system of notation requires some re-examination of deeply ingrained assumptions.

First, we have to distinguish the difference between numbers and the symbols we use to represent numbers. A *number* is a mathematical quantity, usually correlated in electronics to a physical quantity such as voltage, current, or resistance. There are many different types of numbers. Here are just a few types, for example:

Whole Numbers

1, 2, 3, 4, 5, 6, 7, 8, 9 . . .

Integers

-4, -3, -2, -1, 0, 1, 2, 3, 4 . . .

Irrational Numbers

π (approx. 3.1415927), e (approx. 2.718281828),

square root of any prime

Real Numbers

(All one-dimensional numerical values, negative and positive, including zero, whole, integer, and irrational numbers)

Complex Numbers

3 - j4 , 34.5 ∠ 20°

Different types of numbers find different application in the physical world. Whole numbers work well for counting discrete objects, such as the number of resistors in a circuit. Integers are needed when negative equivalents of whole numbers are required. Irrational numbers are numbers that cannot be exactly expressed as the ratio of two integers, and the ratio of a perfect circle's circumference to its diameter (π) is a good physical example of this. The non-integer quantities of voltage, current, and resistance that we're used to dealing with in DC circuits can be expressed as real numbers, in either fractional or decimal form. For AC circuit analysis, however, real numbers fail to capture the dual essence of magnitude and phase angle, and so we turn to the use of complex numbers in either rectangular or polar form.

If we are to use numbers to understand processes in the physical world, make scientific predictions, or balance our checkbooks, we must have a way of symbolically denoting them. In other words, we may know how much money we have in our checking account, but to keep a record of it we need to have some system worked out to symbolize that quantity on paper, or in some other kind of form for record-keeping and tracking. There are two basic ways we can do this: analog and digital. With analog representation, the quantity is symbolized in a way that is infinitely divisible. With digital representation, the quantity is symbolized in a way that is discretely packaged.

You're probably already familiar with an analog representation of money and didn't realize it for what it was. Have you ever seen a fund-raising poster made with a picture of a thermometer on it, where the height of the red column indicated the amount of money collected for the cause? The more money collected, the taller the column of red ink on the poster. The analog representation of the number is depicted in Figure 4.4.

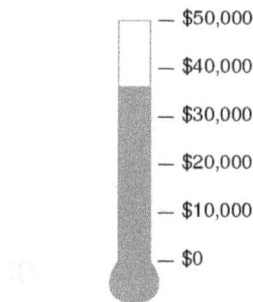

Figure 4.4: Analog Representation of the Number.

This is an example of an analog representation of a number. There is no real limit to how finely divided the height of that column can be made to symbolize the amount of money in the account. Changing the height of that column is something that can be done without changing the essential nature of what it is. Length is a physical quantity that can be divided as small as you would like, with no practical limit. The slide rule is a mechanical device that uses the very same physical quantity—length—to represent number, and to help perform arithmetical operations with two or more numbers at a time. It, too, is an analog device. On the other hand, a digital representation of that same monetary figure, written with standard symbols (sometimes called ciphers), looks like $35,955.38

Unlike the "thermometer" poster with its red column, those symbolic characters above cannot be finely divided: that particular combination of ciphers stand for one quantity and one quantity only. If more money is added to the account (+ $40.12), different symbols must be used to represent the new balance ($35,995.50), or at least the same symbols arranged in different patterns. This is an example of digital representation. The counterpart to the slide rule (analog) is also a digital device: the abacus, with beads that are moved back and forth on rods to symbolize numerical quantities (Figures 4.5 and 4.6):

Slide rule *(an analog device)*

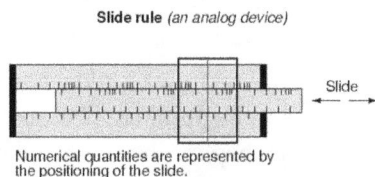

Slide

Numerical quantities are represented by
the positioning of the slide.

Figure 4.5: Digital Representation.

Abacus *(a digital device)*

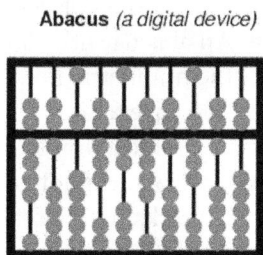

Figure 4.6: Numerical Quantitites are Represented by the Discrete Positions of the Beads.

Methods of Numerical Representation

☆ Analog

☆ Digital

Interpretation of numerical symbols is something we tend to take for granted because it has been taught to us for many years. However, if you were to try to communicate a quantity of something to a person ignorant of decimal numerals, that person could still understand the simple thermometer chart.

The infinitely divisible vs. discrete and precision comparisons are really flip-sides of the same coin. The fact that digital representation is composed of individual, discrete symbols (decimal digits and abacus beads) necessarily means that it will be able to symbolize quantities in precise steps. On the other hand, an analog representation (such as a slide rule's length) is not composed of individual steps, but rather a continuous range of motion. The ability for a slide rule to characterize a numerical quantity to infinite resolution is a trade-off for imprecision. If a slide rule is bumped, an error will be introduced into the representation of the number that was "entered" into it. However, an abacus must be bumped much harder before its beads are completely dislodged from their places (sufficient to represent a different number).

Divisibility of analog versus digital representation can be further illuminated by talking about the representation of irrational numbers. Numbers such as π are called irrational because they cannot be exactly expressed as the fraction of integers, or whole numbers. Although you might have learned in the past that the fraction 22/7 can be used for π in calculations, this is just an approximation. The actual number "pi" cannot be exactly expressed by any finite, or limited, number of decimal places. The digits of π go on forever 3.1415926535897932384......

It is possible, at least theoretically, to set a slide rule (or even a thermometer column) so as to perfectly represent the number π, because analog symbols have no minimum limit to the degree that they can be increased or decreased. If my slide rule shows a figure of 3.141593 instead of 3.141592654, I can bump the slide just a bit more (or less) to get it closer yet. However, with digital representation, such as with an abacus, I would need additional rods (placeholders, or digits) to represent π to further degrees of precision. An abacus with 10 rods simply cannot represent any more than 10 digits worth of the number π, no matter how I set the beads. To perfectly represent π, an abacus would have to have an infinite number of beads and rods! The tradeoff, of course, is the practical limitation to adjusting, and reading analog symbols. Practically speaking, one cannot read a slide rule's scale to the 10th digit of precision, because the marks on the scale are too coarse and human vision is too limited. An abacus, on the other hand, can be set and read with no interpretational errors at all.

Furthermore, analog symbols require some kind of standard by which they can be compared for precise interpretation. Slide rules have markings printed along the length of the slides to translate length into standard quantities. Even the thermometer chart has numerals written along its height to show how much money (in dollars) the red column represents for any given amount of height. Imagine if we all tried to communicate simple numbers to each other by spacing our hands apart varying distances. The number 1 might be signified by holding our hands 1 inch apart, the number 2 with 2 inches, and so on. If someone held their hands 17 inches apart to represent the number 17, would everyone around them be able to immediately and accurately interpret that distance as 17? Some would guess short (15 or 16) and some would guess long (18 or 19). Of course, fishermen who brag about their catches don't mind overestimations in quantity!

Perhaps this is why people have generally settled upon digital symbols for representing numbers, especially whole numbers and integers, which find the most application in everyday life. Using the fingers on our hands, we have a ready means of symbolizing integers from 0 to 10. We can make hash marks on paper, wood, or stone to represent the same quantities quite easily:

$$5 \ + \ 5 \ + \ 3 \ = 13$$

For large numbers, though, the "hash mark" numeration system is too inefficient.

Systems of Numeration

The Romans devised a system that was a substantial improvement over hash marks because it used a variety of symbols (or *ciphers*) to represent increasingly large quantities. The notation for 1 is the capital letter I. The notation for 5 is the capital letter V. Other ciphers possesses increasing values:

X = 10

L = 50

C = 100

D = 500

M = 1000

If a cipher is accompanied by another cipher of equal or lesser value to the immediate right of it, with no ciphers greater than that other cipher to the right of that other cipher, that other cipher's value is added to the total quantity. Thus, VIII symbolizes the number 8, and CLVII symbolizes the number 157. On the other hand, if a cipher is accompanied by another cipher of lesser value to the immediate left, that other cipher's value is *subtracted* from the first. Therefore, IV symbolizes the number 4 (V minus I), and CM symbolizes the number 900 (M minus C). You might have noticed that ending credit sequences for most motion pictures contain a notice for the date of production, in Roman numerals. For the year 1987, it would read MCMLXXXVII. Let's break this numeral down into its constituent parts, from left to right:

M = 1000

+

CM = 900

+

L = 50

+

XXX = 30

+

V = 5

+

II = 2

Aren't you glad we don't use this system of numeration? Large numbers are very difficult to denote this way, and the left vs. right/subtraction vs. addition of values can be very confusing, too. Another major problem with this system is that there is no provision for representing the number zero or negative numbers, both very important concepts in mathematics. Roman culture, however, was more pragmatic with respect to mathematics than most, choosing only to develop their numeration system as far as it was necessary for use in daily life.

We owe one of the most important ideas in numeration to the ancient Babylonians, who were the first (as far as we know) to develop the concept of cipher position, or place value, in representing larger numbers. Instead of inventing new ciphers to represent larger numbers, as the Romans did, they re-used the same ciphers, placing them in different positions from right to left. Our own decimal numeration system uses this concept, with only ten ciphers (0, 1, 2, 3, 4, 5, 6, 7, 8, and 9) used in "weighted" positions to represent very large and very small numbers.

Each cipher represents an integer quantity, and each place from right to left in the notation represents a multiplying constant, or *weight*, for each integer quantity. For example, if we see the decimal notation "1206", we have known that this may be broken down into its constituent weight-products as such:

1206 = 1000 + 200 + 6

$1206 = (1 \times 1000) + (2 \times 100) + (0 \times 10) + (6 \times 1)$

Each cipher is called a *digit* in the decimal numeration system, and each weight, or *place value*, is ten times that of the one to the immediate right. So, we have a *ones* place, a *tens* place, a *hundred* place, a *thousand* place, and so on, working from right to left.

Right about now, you're probably wondering why I'm laboring to describe the obvious. Who needs to be told how decimal numeration works after you've studied math as advanced as algebra and trigonometry? The reason is to better understand other numeration systems, by first knowing the how's and why's of the one you're already used to.

The decimal numeration system uses ten ciphers, and place-weights that are multiples of ten. What if we made a numeration system with the same strategy of weighted places, except with fewer or more ciphers?

The binary numeration system is such a system. Instead of ten different cipher symbols, with each weight constant being ten times the one before it, we only have *two* cipher symbols, and each weight constant is *twice* as much as the one before it. The two allowable cipher symbols for the binary system of numeration are "1" and "0," and these ciphers are arranged right-to-left in doubling values of weight. The rightmost place is the *ones* place, just as with decimal notation. Proceeding to the left, we have the *twos* place, the *fours* place, the *eights* place, the *sixteen's* place, and so on. For example, the following binary number can be expressed, just like the decimal number 1206, as a sum of each cipher value times its respective weight constant:

$$11010 = 2 + 8 + 16 = 26$$

$$11010 = (1 \times 16) + (1 \times 8) + (0 \times 4) + (1 \times 2) + (0 \times 1)$$

This can get quite confusing, as I've written a number with binary numeration (11010), and then shown its place values and total in standard, decimal numeration form (16 + 8 + 2 = 26). In the above example, we're mixing two different kinds of numerical notation. To avoid unnecessary confusion, we have to denote which form of numeration we're using when we write (or type!). Typically, this is done in the subscript form, with a "2" for binary and a "10" for decimal, so the binary number 11010_2 is equal to the decimal number 26_{10}.

The subscripts are not mathematical operation symbols like superscripts (exponents) are. All they do is indicate what system of numeration we're using when we write these symbols for other people to read. If you see "3_{10}", this entire means is the number three written using *decimal* numeration. However, if you see "3^{10}", this means something completely different: three to the tenth power (59,049). As usual, if no subscript is shown, the cipher(s) are assumed to be representing a decimal number.

Commonly, the number of cipher types (and therefore, the place-value multiplier) used in a numeration system is called that system's *base*. Binary is referred to as "base two" numeration, and decimal as "base ten." Additionally, we refer to each cipher position in binary as a *bit* rather than the familiar word *digit* used in the decimal system.

Now, why would anyone use binary numeration? The decimal system, with its ten ciphers, makes a lot of sense, being that we have ten fingers on which to count between our two hands. (It is interesting that some ancient central American cultures used numeration systems with a base of twenty. Presumably, they used both fingers and toes to count!!). But the primary reason that the binary numeration system is used in modern electronic computers is that of the ease of representing two cipher states (0 and 1) electronically. With relatively simple circuitry, we can perform mathematical operations on binary numbers by representing each bit of the numbers by a circuit which is either on (current) or off (no current). Just like the abacus with each rod representing another decimal digit, we simply add more circuits to give us more bits to symbolize larger numbers. Binary numeration also lends itself well to the storage and retrieval of numerical information: on magnetic tape (spots of iron oxide on the tape either being magnetized for a binary "1" or demagnetized for a binary "0"), optical disks (a laser-burned pit in the aluminum foil representing a binary "1" and an unburned spot representing a binary "0"), or a variety of other media types.

Before we go on to learning exactly how all this is done in digital circuitry, we need to become more familiar with binary and other associated systems of numeration.

Decimal Versus Binary Numeration

Let's count from zero to twenty using four different kinds of numeration systems: hash marks, Roman numerals, decimal, and binary:

System:	Hash Marks	Roman	Decimal	Binary
Zero	n/a	n/a	0	0
One	|	I	1	1
Two	||	II	2	10
Three	|||	III	3	11
Four	||||	IV	4	100
Five	/|||/	V	5	101
Six	/|||/ |	VI	6	110
Seven	/|||/ ||	VII	7	111
Eight	/|||/ |||	VIII	8	1000
Nine	/|||/ ||||	IX	9	1001
Ten	/|||/ /|||/	X	10	1010
Eleven	/|||/ /|||/ |	XI	11	1011
Twelve	/|||/ /|||/ ||	XII	12	1100
Thirteen	/|||/ /|||/ |||	XIII	13	1101
Fourteen	/|||/ /|||/ ||||	XIV	14	1110
Fifteen	/|||/ /|||/ /|||/	XV	15	1111
Sixteen	/|||/ /|||/ /|||/ |	XVI	16	10000
Seventeen	/|||/ /|||/ /|||/ ||	XVII	17	10001

Eighteen	/			/ /			/ /			/				XVIII	18	10010
Nineteen	/			/ /			/ /			/					XIX	19	10011
Twenty	/			/ /			/ /			/ /			/	XX	20	10100

Neither hash marks nor the Roman system is very practical for symbolizing large numbers. Obviously, place-weighted systems such as decimal and binary are more efficient for the task. Notice, though, how much shorter decimal notation is over binary notation, for the same number of quantities. What takes five bits in binary notation only takes two digits in decimal notation.

This raises an interesting question regarding different numeration systems: how large of a number can be represented with a limited number of cipher positions or places? With the crude hash-mark system, the number of places IS the largest number that can be represented, since one hash mark "place" is required for every integer step. For place-weighted systems of numeration, however, the answer is found by taking the base of the numeration system (10 for decimal, 2 for binary) and raising it to the power of the number of places. For example, 5 digits in a decimal numeration system can represent 100,000 different integer number values, from 0 to 99,999 (10 to the 5th power = 100,000). 8 bits in a binary numeration system can represent 256 different integer number values, from 0 to 11111111 (binary), or 0 to 255 (decimal), because 2 to the 8th power equals 256. With each additional place position to the number field, the capacity for representing numbers increases by a factor of the base (10 for decimal, 2 for binary).

An interesting footnote for this topic is one of the first electronic digital computers, the Eniac. The designers of the Eniac chose to represent numbers in decimal form, digitally, using a series of circuits called "ring counters" instead of just going with the binary numeration system, in an effort to minimize the number of circuits required to represent and calculate very large numbers. This approach turned out to be counter-productive, and virtually all digital computers since then have been purely binary in design.

To convert a number in binary numeration to its equivalent in decimal form, all you have to do is calculate the sum of all the products of bits with their respective place-weight constants. To illustrate:

Convert 11001101_2 to decimal form:

bits =	1 1 0 0 1 1 0 1
	- - - - - - - -
weight =	1 6 3 1 8 4 2 1
(in decimal	2 4 2 6
notation)	8

The bit on the far right side is called the Least Significant Bit (LSB), because it stands in the place of the lowest weight (the one's place). The bit on the far left side is called the Most Significant Bit (MSB), because it stands in the place of the highest weight (the one hundred twenty-eight's place). Remember, a bit value of "1" means that the respective place weight gets added to the total value, and a bit value of "0" means that the respective place weight does not get added to the total value. With the above example, we have:

$$128_{10} + 64_{10} + 8_{10} + 4_{10} + 1_{10} = 205_{10}$$

If we encounter a binary number with a dot (.), called a "binary point" instead of a decimal point, we follow the same procedure, realizing that each place weight to the right of the point is one-half the value of the one to the left of it (just as each place weight to the right of a decimal point is one-tenth the weight of the one to the left of it). For example:

Convert 101.011_2 to decimal form:

```
.

bits =            1 0 1 . 0 1 1

.                 - - - - - - -

weight =          4 2 1 1 1 1
(in decimal       / / /
notation)         2 4 8
```

$$4_{10} + 1_{10} + 0.25_{10} + 0.125_{10} = 5.375_{10}$$

Octal and Hexadecimal Numeration

Because binary numeration requires so many bits to represent relatively small numbers compared to the economy of the decimal system, analyzing the numerical states inside of digital electronic circuitry can be a tedious task. Computer programmers who design sequences of number codes instructing a computer what to do would have a very difficult task if they were forced to work with nothing but long strings of 1's and 0's, the "native language" of any digital circuit. To make it easier for human engineers, technicians, and programmers to "speak" this language of the digital world, other systems of place-weighted numeration have been made which are very easy to convert to and from binary.

One of those numeration systems is called *octal* because it is a place-weighted system with a base of eight. Valid ciphers include the symbols 0, 1, 2, 3, 4, 5, 6, and 7. Each place weight differs from the one next to it by a factor of eight.

Another system is called *hexadecimal* because it is a place-weighted system with a base of sixteen. Valid ciphers include the normal decimal symbols 0, 1, 2, 3, 4, 5, 6, 7, 8, and 9, plus six alphabetical characters A, B, C, D, E, and F, to make a total of sixteen. As you might have guessed already, each place weight differs from the one before it by a factor of sixteen.

Let's count again from zero to twenty using decimal, binary, octal, and hexadecimal to contrast these systems of numeration:

Number	Decimal	Binary	Octal	Hexadecimal
Zero	0	0	0	0
One	1	1	1	1
Two	2	10	2	2
Three	3	11	3	3
Four	4	100	4	4
Five	5	101	5	5
Six	6	110	6	6
Seven	7	111	7	7
Eight	8	1000	10	8
Nine	9	1001	11	9
Ten	10	1010	12	A
Eleven	11	1011	13	B
Twelve	12	1100	14	C
Thirteen	13	1101	15	D
Fourteen	14	1110	16	E
Fifteen	15	1111	17	F
Sixteen	16	10000	20	10
Seventeen	17	10001	21	11
Eighteen	18	10010	22	12
Nineteen	19	10011	23	13
Twenty	20	10100	24	14

Octal and hexadecimal numeration systems would be pointless if not for their ability to be easily converted to and from binary notation. Their primary purpose in being is to serve as a "shorthand" method of denoting a number represented electronically in binary form. Because the bases of octal (eight) and hexadecimal (sixteen) are even multiples of binary's base (two), binary bits can be grouped together and directly converted to or from their respective octal or hexadecimal digits. With octal, the binary bits are grouped in three's (because of $2^3 = 8$), and with hexadecimal, the binary bits are grouped in four's (because of $2^4 = 16$)

Binary to Octal Conversion

Convert 10110111.1_2 to octal:

.

. implied zero   implied zeros

. | ||

. 010 110   111 100

Convert each group of bits   ###   ###   ### . ###

to its octal equivalent: 2   6   7   4

.

Answer: $10110111.1_2 = 267.4_8$

We had to group the bits in three's, from the binary point left, and from the binary point right, adding (implied) zeros as necessary to make complete 3-bit groups. Each octal digit was translated from the 3-bit binary groups. Binary-to-Hexadecimal conversion is much the same:

Binary to Hexadecimal Conversion

Convert 10110111.1_2 to hexadecimal:

.

.   implied zeros

. |||

.   1011 0111 1000

Convert each group of bits   ----   ---- . ----

to its hexadecimal equivalent: B   7   8

.

Answer: $10110111.1_2 = B7.8_{16}$

Here we had to group the bits in four's, from the binary point left, and from the binary point right, adding (implied) zeros as necessary to make complete 4-bit groups:

Likewise, the conversion from either octal or hexadecimal to binary is done by taking each octal or hexadecimal digit and converting it to its equivalent binary (3 or 4 bit) group, then putting all the binary bit groups together.

Incidentally, hexadecimal notation is more popular, because binary bit groupings in digital equipment are common multiples of eight (8, 16, 32, 64, and 128 bit), which are also multiples of 4. Octal, being based on binary bit groups of 3, doesn't work out evenly with those common bit group sizings.

Octal and Hexadecimal to Decimal Conversion

Although the prime intent of octal and hexadecimal numeration systems is for the "shorthand" representation of binary numbers in digital electronics, we sometimes have the need to convert from either of those systems to decimal form. Of course, we could simply convert the hexadecimal or octal format to binary, then convert from binary to decimal, since we already know how to do both, but we can also convert directly.

Because octal is a base-eight numeration system, each place-weight value differs from either adjacent place by a factor of eight. For example, the octal number 245.37 can be broken down into place values as such:

```
octal
digits =      2 4 5 . 3 7
.            - - - - - -
weight =      6 8 1   1 1
(in decimal   4       / /
notation)             8 6
.                 4
```

The decimal value of each octal place-weight times its respective cipher multiplier can be determined as follows:

$$(2 \times 64_{10}) + (4 \times 8_{10}) + (5 \times 1_{10}) + (3 \times 0.125_{10}) + (7 \times 0.015625_{10}) = 165.484375_{10}$$

The technique for converting hexadecimal notation to decimal is the same, except that each successive place-weight changes by a factor of sixteen. Simply denote each digit's weight, multiply each hexadecimal digit value by its respective weight (in decimal form), then add up all the decimal values to get a total. For example, the hexadecimal number $30F.A9_{16}$ can be converted like this:

```
hexadecimal
digits =      3 0 F . A 9
.            - - - - - -
weight =      2 1 1   1 1
(in decimal   5 6     / /
notation)     6       1 2
.                 6 5
.                 6
```

$(3 \times 256_{10}) + (0 \times 16_{10}) + (15 \times 1_{10}) + (10 \times 0.0625_{10}) +$

$(9 \times 0.00390625_{10}) = 783.66015625_{10}$

These basic techniques may be used to convert a numerical notation of any base into decimal form, if you know the value of that numeration system's base.

Conversion from Decimal Numeration

Because octal and hexadecimal numeration systems have bases that are multiples of binary (base 2), conversion back and forth between either hexadecimal or octal and binary is very easy. Also, because we are so familiar with the decimal system, converting binary, octal, or hexadecimal to decimal form is relatively easy (simply add up the products of cipher values and place-weights). However, conversion from decimal to any of these "strange" numeration systems is a different matter.

The method which will probably make the most sense is the "trial-and-fit" method, where you try to "fit" the binary, octal, or hexadecimal notation to the desired value as represented in decimal form. For example, let's say that I wanted to represent the decimal value of 87 in binary form. Let's start by drawing a binary number field, complete with place-weight values: .

```
.          - - - - - - - -
weight =       1 6 3 1 8 4 2 1
(in decimal    2 4 2 6
notation)      8
```

Well, we know that we won't have a "1" bit in the 128's place, because that would immediately give us a value greater than 87. However, since the next weight to the right (64) is less than 87, we know that we must have a "1" there.

```
.     1
.     - - - - - - -    Decimal value so far = 64₁₀
weight =       6 3 1 8 4 2 1
(in decimal    4 2 6
notation)
```

$\text{Decimal value so far} = 64_{10}$

If we were to make the next place to the right a "1" as well, our total value would be $64_{10} + 32_{10}$, or 96_{10}. This is greater than 87_{10}, so we know that this bit must be a "0". If we make the next (16's) place bit equal to "1," this brings our total value to $64_{10} + 16_{10}$, or 80_{10}, which is closer to our desired value (87_{10}) without exceeding it:

```
.          1 0 1
.          - - - - - - -    Decimal value so far = 80₁₀
weight =       6 3 1 8 4 2 1
(in decimal    4 2 6
notation)
```

$\text{Decimal value so far} = 80_{10}$

By continuing in this progression, setting each lesser-weight bit as we need to come up to our desired total value without exceeding it, we will eventually arrive at the correct figure:

.   1 0 1 0 1 1 1

.   - - - - - - -   Decimal value so far $= 87_{10}$

weight =   6 3 1 8 4 2 1

(in decimal   4 2 6

notation)

This trial-and-fit strategy will work with octal and hexadecimal conversions, too. Let's take the same decimal figure, 87_{10}, and convert it to octal numeration:

.   - - -

weight =   6 8 1

(in decimal   4

notation)

If we put a cipher of "1" in the 64's place, we would have a total value of 64_{10} (less than 87_{10}). If we put a cipher of "2" in the 64's place, we would have a total value of 128_{10} (greater than 87_{10}). This tells us that our octal numeration must start with a "1" in the 64's place:

.   1

.   - - -   Decimal value so far $= 64_{10}$

weight =   6 8 1

(in decimal   4

notation)

Now, we need to experiment with cipher values in the 8's place to try and get a total (decimal) value as close to 87 as possible without exceeding it. Trying the first few cipher options, we get:

"1" $= 64_{10} + 8_{10} = 72_{10}$

"2" $= 64_{10} + 16_{10} = 80_{10}$

"3" $= 64_{10} + 24_{10} = 88_{10}$

A cipher value of "3" in the 8's place would put us over the desired total of 87_{10}, so "2" it is!

.   1 2

.   - - -   Decimal value so far $= 80_{10}$

weight =   6 8 1

(in decimal   4

notation)

Now, all we need to make a total of 87 is a cipher of "7" in the 1's place:

.   1 2 7

.   - - -   Decimal value so far = 87_{10}

weight =   6 8 1

(in decimal   4

notation)

Of course, if you were paying attention during the last section on octal/binary conversions, you will realize that we can take the binary representation of (decimal) 87_{10}, which we previously determined to be 1010111_2, and easily convert from that to octal to check our work:

. Implied zeros

. ||

.   001 010 111 Binary

.   — --- —

.   1   2   7   Octal

.

Answer: $1010111_2 = 127_8$

Can we do decimal-to-hexadecimal conversion the same way? Sure, but who would want to? This method is simple to understand, but laborious to carry out. There is another way to do these conversions, which is essentially the same (mathematically), but easier to accomplish.

This other method uses repeated cycles of division (using decimal notation) to break the decimal numeration down into multiples of binary, octal, or hexadecimal place-weight values. In the first cycle of division, we take the original decimal number and divide it by the base of the numeration system that we're converting to (binary = 2 octal = 8, hex = 16). Then, we take the whole-number portion of division result (quotient) and divide it by the base value again, and so on, until we end up with a quotient of less than 1. The binary, octal, or hexadecimal digits are determined by the "remainders" left over by each division step. Let's see how this works for binary, with the decimal example of 87_{10}:

. 87   Divide 87 by 2, to get a quotient of 43.5

. — = 43.5   Division "remainder" = 1, or the < 1 portion

. 2   of the quotient times the divisor (0.5×2)

.

$$\frac{43}{2} = 21.5$$ Take the whole-number portion of 43.5 (43) and divide it by 2 to get 21.5, or 21 with a remainder of 1

$$\frac{21}{2} = 10.5$$ And so on ... remainder = 1 (0.5×2)

$$\frac{10}{2} = 5.0$$ And so on ... remainder = 0

$$\frac{5}{2} = 2.5$$ And so on ... remainder = 1 (0.5×2)

$$\frac{2}{2} = 1.0$$ And so on ... remainder = 0

$$\frac{1}{2} = 0.5$$ until we get a quotient of less than 1
remainder = 1 (0.5×2)

The binary bits are assembled from the remainders of the successive division steps, beginning with the LSB and proceeding to the MSB. In this case, we arrive at a binary notation of 1010111_2. When we divide by 2, we will always get a quotient ending with either ".0" or ".5", *i.e.* a remainder of either 0 or 1. As was said before, this repeat-division technique for conversion will work for numeration systems other than binary. If we were to perform successive divisions using a different number, such as 8 for conversion to octal, we will necessarily get remainders between 0 and 7. Let's try this with the same decimal number, 87_{10}:

$$\frac{87}{8} = 10.875$$ Divide 87 by 8, to get a quotient of 10.875
Division "remainder" = 7, or the < 1 portion of the quotient times the divisor ($.875 \times 8$)

. 10

. — = 1.25   Remainder = 2

. 8

.

. 1

. — = 0.125   Quotient is less than 1, so we'll stop here.

.   8 Remainder = 1

.

. RESULT: $87_{10} = 127_8$

We can use a similar technique for converting numeration systems dealing with quantities less than 1, as well. For converting a decimal number less than 1 into binary, octal, or hexadecimal, we use repeated multiplication, taking the integer portion of the product in each step as the next digit of our converted number. Let's use the decimal number 0.8125_{10} as an example, converting to binary:

. $0.8125 \times 2 = 1.625$ Integer portion of product = 1

.

. $0.625 \times 2 = 1.25$ Take < 1 portion of product and remultiply

.   Integer portion of product = 1

.

. $0.25 \times 2 = 0.5$ Integer portion of product = 0

.

. $0.5 \times 2 = 1.0$ Integer portion of product = 1

.   Stop when product is a pure integer

.   (ends with .0)

.

. RESULT: $0.8125_{10} = 0.1101_2$

As with the repeat-division process for integers, each step gives us the next digit (or bit) further away from the "point." With integer (division), we worked from the LSB to the MSB (right-to-left), but with repeated multiplication, we worked from the left to the right. To convert a decimal number greater than 1, with a < 1 component, we must use *both* techniques, one at a time. Take the decimal example of 54.40625_{10}, converting to binary:

REPEATED DIVISION FOR INTEGER PORTION:

.

. 54

. — = 27.0   Remainder = 0

. 2

.

. 27

. — = 13.5   Remainder = 1 (0.5 × 2)

. 2

.

. 13

. — = 6.5   Remainder = 1 (0.5 × 2)

. 2

.

. 6

. — = 3.0   Remainder = 0

. 2

.

. 3

. — = 1.5   Remainder = 1 (0.5 × 2)

. 2

.

. 1

. — = 0.5   Remainder = 1 (0.5 × 2)

. 2

.

PARTIAL ANSWER: $54_{10} = 110110_2$

REPEATED MULTIPLICATION FOR THE < 1 PORTION:

.

. $0.40625 \times 2 = 0.8125$ Integer portion of product = 0

.

. $0.8125 \times 2 = 1.625$ Integer portion of product = 1

· 0.625 × 2 = 1.25   Integer portion of product = 1

· 0.25 × 2 = 0.5   Integer portion of product = 0

· 0.5 × 2 = 1.0   Integer portion of product = 1

· PARTIAL ANSWER: $0.40625_{10} = 0.01101_2$

· COMPLETE ANSWER: $54_{10} + 0.40625_{10} = 54.40625_{10}$

· $110110_2 + 0.01101_2 = 110110.01101_2$

Binary-to-Decimal Conversion

Any binary number can be converted to its decimal equivalent simply by summing together the weights of the various positions in the binary number which contain a 1.

$$1\,1\,0\,1\,1_2 \text{ (binary)}$$
$$2^4+2^3+0+2^1+2^0 = 16+8+0+2+1$$
$$= 27_{10} \text{ (decimal)}$$

and

$$1\,0\,1\,1\,0\,1\,0\,1_2 \text{ (binary)}$$
$$2^7+0+2^5+2^4+0+2^2+0+2^0 = 128+0+32+16+0+4+0+1$$
$$= 181_{10} \text{ (decimal)}$$

You should notice the method finds the weights (*i.e.*, powers of 2) for each bit position that contains a 1, and then to add them up.

Decimal-To-Binary Conversion

There are 2 methods:

(A) Revere of Binary-To-Digital Method

$$45_{10} = 32 + 0 + 8 + 4 + 0 + 1$$
$$= 2^5+0+2^3+2^2+0+2^0$$
$$= 1\,0\,1\,1\,0\,1_2$$

(B) Repeat Division

This method uses repeated division by 2. Eg. convert 25_{10} to binary

25/ 2	= 12+ remainder of 1	1 (Least Significant Bit)
12/ 2	= 6 + remainder of 0	0
6 / 2	= 3 + remainder of 0	0
3 / 2	= 1 + remainder of 1	1
1 / 2	= 0 + remainder of *1*	*1 (Most Significant Bit)*
Result	25_{10} =	*1 1 0 0 1* $_2$

The Flowchart for repeated-division method is as follow:

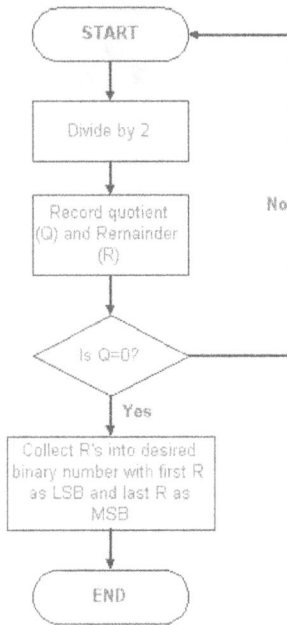

Octal Number System

The octal number system has a base of eight, meaning that it has eight possible digits: 0,1,2,3,4,5,6,7.

8^3	8^2	8^1	8^0		8^{-1}	8^{-2}	8^{-3}
= 512	= 64	= 8	= 1	.	= 1/8	= 1/64	= 1/512
Most Significant Digit				Octal point			Least Significant Digit

Octal to Decimal Conversion

Eg. $24.6_8 = 2 \times (8^1) + 4 \times (8^0) + 6 \times (8^{-1}) = 20.75_{10}$

Binary-To-Octal / Octal-To-Binary Conversion

Octal Digit	0	1	2	3	4	5	6	7
Binary Equivalent	000	001	010	011	100	101	110	111

Each Octal digit is represented by three bits of binary digit.

Eg. $100\ 111\ 010_2 = (100)\ (111)\ (010)_2 = 4\ 7\ 2_8$

Repeat Division

This method uses repeated division by 8. Eg. convert 177_{10} to octal and binary:

177/8	= 22+ remainder of 1	1 (Least Significant Bit)
22/ 8	= 2 + remainder of 6	6
2 / 8	= 0 + remainder of 2	2 *(Most Significant Bit)*
Result	$177_{10} =$	261_8
	Convert to Binary	$= 010110001_2$

Decimal-To-Binary Conversion

There are 2 methods:

(A) Reverse of Binary-To-Digital Method

$$45_{10} = 32 + 0 + 8 + 4 + 0 + 1$$
$$= 2^5 + 0 + 2^3 + 2^2 + 0 + 2^0$$
$$= 1\ 0\ 1\ 1\ 0\ 1_2$$

(B) Repeat Division

This method uses repeated division by 2. Eg. convert 25_{10} to binary

25/ 2	= 12+ remainder of 1	1 (Least Significant Bit)
12/ 2	= 6 + remainder of 0	0
6 / 2	= 3 + remainder of 0	0
3 / 2	= 1 + remainder of 1	1
1 / 2	= 0 + remainder of 1	1 (Most Significant Bit)
Result	$25_{10} =$	$1\ 1\ 0\ 0\ 1_2$

The Flow chart for repeated-division method is as follow:

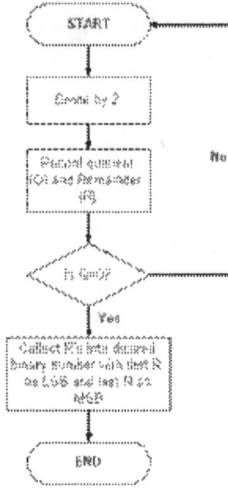

Memory Circuits

Introduction

The memory inside a computer is organized in a matrix of rows and columns. Each word of memory is uniquely identified by the combined row and column values - known as the address. This arrangement resembles a set of pigeon holes used to sort post. Each item of a post has an assigned address, which indicates the pigeonhole into which it is to be placed:

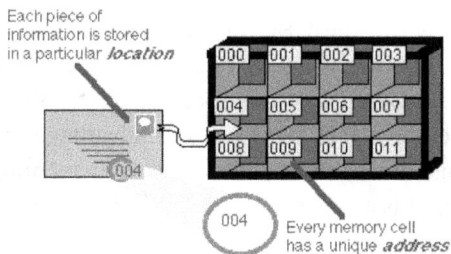

For instance, to read the value stored at a memory location 4, the following steps of actions are performed:

(i) Set the memory address value to 4 to indicate that we are going to deal with the 4th memory location

(ii) Set the memory control logic to READ (enables the memory)

(iii) The data register returns the value at the location specified by the memory address.

(iv) DISABLE the memory control logic

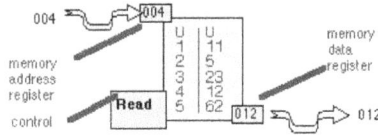

Memory Hierarchy

As often is the case, we utilize a number of logical models of our memory system, depending on the point we want to make. The simplest view of memory is that of a **monolithic linear memory**; specifically, a memory fabricated as a single unit (**monolithic**) that is organized as a single dimensioned array (**linear**). This is satisfactory as a logical model, but it ignores very many issues of considerable importance.

Consider a memory in which an M–bit word is the smallest addressable unit. For simplicity, we assume that the memory contains $N = 2^K$ words and that the address space is also $N = 2^K$. The memory can be viewed as a one-dimensional array, declared something like Memory: Array $[0 .. (N - 1)]$ of M–bit word.

The **monolithic view** of the memory is shown in the Figure 5.1.

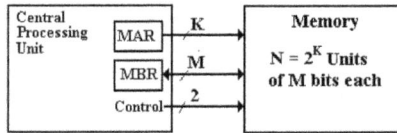

Figure 5.1: Monolithic View of Computer Memory.

In this monolithic view, the CPU provides K address bits to access $N = 2^K$ memory entries, each of which has M bits, and at least two control signals to manage memory. The **linear view** of memory is a way to think logically about the organization of the memory. This view has the advantage of being rather simple but has the disadvantage of describing accurately only technologies that have long been obsolete. However, it is a consistent model that is worth mention (Figure 5.2).

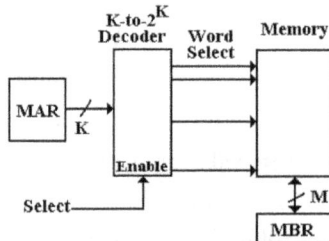

Figure 5.2: Linear Model.

There are two problems with the above model, a minor nuisance and a "show–stopper". The minor problem is the speed of the memory; its access time will be exactly that of plain variety DRAM (**d**ynamic **r**andom **a**ccess **m**emory), which is about 50 nanoseconds. We must have better performance than that, so we go to other memory organizations.

The "show–stopper" problem is the design of the memory decoder. Consider two examples for common memory sizes: 1MB (2^{20} bytes) and 4GB (2^{32} bytes) in a byte-oriented memory.

A 1MB memory would use a 20–to–1,048,576 decoder, as $2^{20} = 1,048,576$.

A 4GB memory would use a 32–to–4,294,967,296 decoder, as $2^{32} = 4,294,967,296$.

Neither of these decoders can be manufactured at acceptable cost using current technology. One should note that since other designs offer additional benefits, there is no motivation to undertake such a design task.

We now examine two design choices that produce easy-to-manufacture solutions that offer acceptable performance at reasonable price. The first is the memory hierarchy, using various levels of cache memory, offering faster access to main memory. Before discussing the ideas behind cache memory, we must first speak of two technologies for fabricating memory.

RAM

Random Access Memory (RAM) is used for storing most programs and data in a computer (Figure 5.3). Most computers also use some Read Only Memory (ROM) or Erasable-Programmable Read Only Memory (EPROM). This stores programs (or libraries of procedures) which do not need to be changed. (This usually includes the program that is executed when the computer is first turned on, and the procedures required to access permanently connected peripherals - such as the keyboard, display, disk drive).

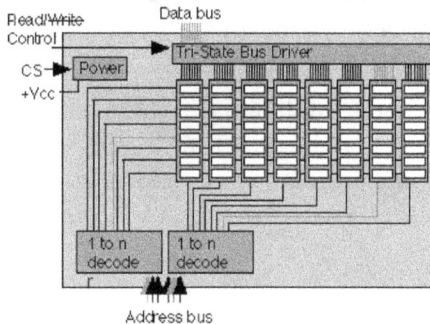

Figure 5.3: Random Access Memory.

SRAM (Static RAM) and DRAM (Dynamic RAM)

We now discuss technologies used to store binary information. The first topic is to make a list of requirements for devices used to implement binary memory.

1. Two well defined and distinct states.
2. The device must be able to switch states reliably.
3. The probability of a spontaneous state transition must be extremely low.
4. State switching must be as fast as possible.
5. The device must be small and cheap so that large capacity memories are practical.

There are a number of memory technologies that were developed in the last half of the twentieth century. Most of these are now obsolete. There are three that are worth mention:

1. Core Memory (now obsolete, but new designs may be introduced soon)
2. Static RAM
3. Dynamic RAM

Core Memory

This was a major advance when it was introduced in 1952, first used on the MIT Whirlwind. The basic memory element is a torus of magnetic material. This torus can contain a magnetic field in one of two directions. These two distinct directions allow for a two-state device, as required to store a binary number.

One aspect of magnetic core memory remains with us – the frequent use of the term "core memory" as a synonym for the computer's main memory. In discussing the next two technologies, we must make two definitions. These are memory access time and memory cycle time. The measure of real interest is the access time. The unit for measuring memory times is the nanosecond, equal to one billionth of a second.

Memory access time is the time required for the memory to access the data; specifically, it is the time between the instant that the memory address is stable in the MAR and the data are available in the MBR.

Memory cycle time is the minimum time between two independent memory accesses. It should be clear that the cycle time is at least as great as the access time because the memory cannot process an independent access while it is in the process of placing data in the MBR.

Static RAM

Static RAM (SRAM) is a memory technology based on flip-flops. SRAM has an access time of 2 – 10 nanoseconds. From a logical view, all of the main memory can be viewed as fabricated from SRAM, although such a memory would be unrealistically expensive.

Dynamic RAM

Dynamic RAM (DRAM) is a memory technology based on capacitors – circuit elements that store electronic charge. Dynamic RAM is cheaper than static RAM and can be packed more densely on a computer chip, thus allowing larger capacity memories. DRAM has an access time in the order of 60 – 100 nanoseconds, slower than SRAM.

Our goal in designing a memory would be to create a memory unit with the performance of SRAM, but the cost and packaging density of DRAM. Fortunately, this is possible by use of a design strategy called "cache memory", in which a fast SRAM memory fronts for a larger and slower DRAM memory. This design has been found to be useful due to a property of computer program execution called "**program locality**". It is a fact that if a program accesses a memory location, it is likely both to access that location again and to access locations with similar addresses. It is program locality that allows cache memory to work.

Static RAM is composed of D-Type Flip-Flops, and is extremely fast, however it is also expensive. It is therefore usually reserved for applications requiring a high speed (such as graphics display memory or cache memory. The main computer memory is usually formed from Dynamic RAM (DRAM), which uses an array of capacitor storage elements. Although slower than SRAM it is also much cheaper.

ROM

There is a type of memory that stores data without electrical current; it is the **ROM** (*Read Only Memory*) or is sometimes called *non-volatile memory* as it is not erased when the system is switched off. This type of memory lets you stored the data needed to start up the computer. Indeed, this information cannot be stored on the hard disk since the disk parameters (vital for its initialization) are part of these data which are essential for booting.

Different *ROM*-type memories contain these essential start-up data, *i.e.*:

 ☆ The BIOS is a programme for controlling the system's main input-output interfaces, hence the name *BIOS ROM* which is sometimes given to the read-only memory chip of the motherboard which hosts it.

 ☆ The **bootstrap loader**: a programme for loading (random access) memory into the operating system and launching it. This generally seeks the operating system on the floppy drive then on the hard disk, which allows the operating system to be launched from a system floppy disk in the event of malfunction of the system installed on the hard disk.

 ☆ The **CMOS Setup** is the screen displayed when the computer starts up and which is used to amend the system parameters (often wrongly referred to as *BIOS*).

☆ The **Power-On Self Test** (*POST*), a programme that runs automatically when the system is booted, thus allowing the system to be tested (this is why the system "counts" the RAM at start-up).

Given that ROM is much slower than RAM memories (access time for a ROM is around 150 ns whereas for SDRAM it is around 10 ns), the instructions given in the ROM are sometimes copied to the RAM at start-up; this is known as *shadowing*, though is usually referred to as *shadow memory*).

Types of ROM

ROM memories have gradually evolved from *fixed read-only memories* to memories that can be programmed and then re-programmed.

ROM

The first ROMs were made using a procedure that directly writes the binary data in a silicon plate using a mask. This procedure is now obsolete.

PROM

PROM (*Programmable Read Only Memory*) memories were developed at the end of the 70s by a company called *Texas Instruments*. These memories are chips comprising thousands of fuses (or diodes) that can be "burnt" using a device called a "*ROM programmer*", applying high voltage (12V) to the memory boxes to be marked. The fuses thus burnt correspond to 0 and the others to 1.

EPROM

EPROM (*Erasable Programmable Read Only Memory*) memories are PROMs that can be deleted. These chips have a glass panel that lets ultra-violet rays through. When the chip is subjected to ultra-violet rays with a certain wavelength, the fuses are reconstituted, meaning that all the memory bits return to 1. This is why this type of PROM is called *erasable*.

EEPROM

EEPROM (*Electrically Erasable Read Only Memory*) memories are also erasable PROMs, but unlike EPROMs, they can be erased by a simple electric current, meaning that they can be erased even when they are in a position in the computer.

There is a variant of these memories known as **flash memories** (also *Flash ROM* or *Flash EPROM*). Unlike the classic EEPROMs that use 2 to 3 transistors for each bit to be memorized, the EPROM Flash uses only one transistor. Moreover, the EEPROM may be written and read word by word, while the Flash can be erased only in pages (the size of the pages decreases constantly).

Unit 6

Programmable Logic Devices

PLA

Introduction

One way to design a combinational logic circuit it to get gates and connect them with wires. One disadvantage with this way of designing circuits is its lack of portability. You can now get chips called PLA (programmable logic arrays) and "program" them to implement Boolean functions. I'll explain what it means to program a PLA. Fortunately, a PLA is quite simple to learn and produces nice neat circuits too (Figure 6.1).

Starting Out

The first part of a PLA looks like:

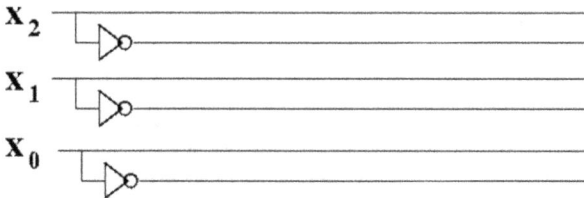

Figure 6.1: Programmable Logic Arrays.

Each variable is hooked to a wire, and to a wire with a NOT gate. So the top wire is x_2 and the one just below is its negation, $\backslash x_2$. Then there's x_1 and just below it, its negation, $\backslash x_1$. The next part is to draw a vertical wire with an AND gate.

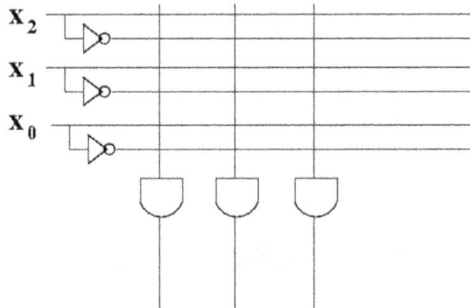

Let's try to implement a truth table with a PLA.

x_2	x_1	x_0	z_1	z_0
0	0	0	0	0
0	0	1	1	0
0	1	0	0	0
0	1	1	1	0
1	0	0	1	1
1	0	1	0	0
1	1	0	0	0
1	1	1	0	1

Each of the vertical lines with an AND gate corresponds to a minterm. For example, the first AND gate (on the left) is the minterm: $\backslash x_2 \backslash x_1 x_0$.

The second AND gate (from the left) is the minterm: $\backslash x_2 x_1 x_0$.

The third AND gate (from the left) is the minterm: $x_2 \backslash x_1 \backslash x_0$.

I've added a fourth AND gate which is the minterm: $x_2 x_1 x_0$. The first three minterms are used to implement z_1. The third and fourth minterm are used to implement z_0. This is how the PLA looks after we have all four minterms (Figure 6.2).

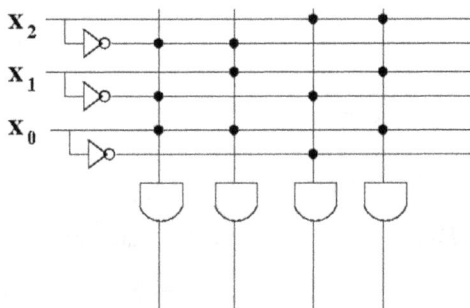

Figure 6.2: PLA with Minterms.

For each connection (shown with a black dot), there's really a separate wire. We draw one wire just to make it look neat (Figure 6.3).

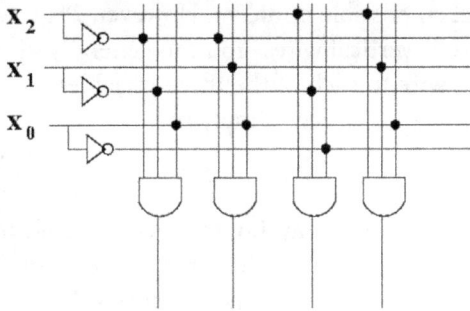

Figure 6.3: PLA.

The vertical wires are called the AND plane. We often leave out the AND gates to make it even easier to draw. We then add OR gates using horizontal wires, to connect the minterms together.

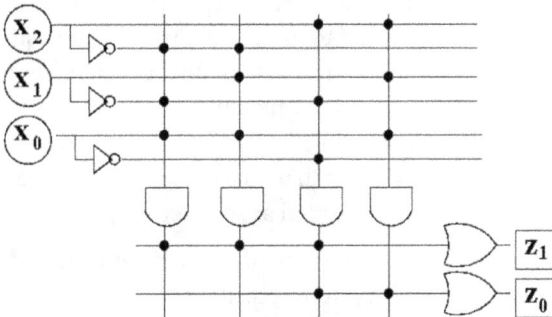

Again, a single wire into the OR gate is really 4 wires. We use the same simplification to make it easier to read. The horizontal wires make up the OR plane. This is how the PLA looks when we leave out the AND gates and the OR gates. It's not that the AND gates and OR gates aren't there---they are, but they've been left out to make the PLA even easier to draw.

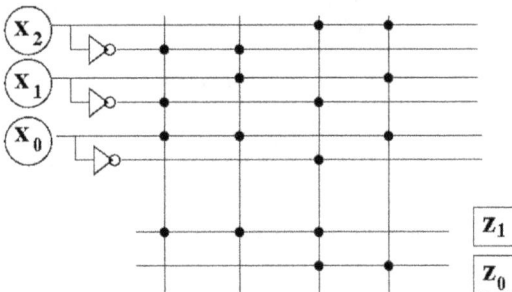

Minterms

Given **n** variables, it would seem necessary to have 2^n vertical wires (for the AND gates), one for each possible minterm. However, 2^n grows VERY quickly. So, sometimes there aren't 2^n vertical wires. You can generally get around the problem by not connecting the wire to each of the three variables. For example, you could just have a product term $x_2 \backslash x_0$ or even simply $\backslash x_1$.

PAL

The term Programmable Array Logic (PAL) is used to describe a family of programmable logic device semiconductors used to implement logic functions in digital circuits introduced by Monolithic Memories, Inc. (MMI) in March 1978.

The PAL device is a special case of PLA which has a programmable AND array and a fixed OR array. The basic structure of Rom is same as PLA. It is cheap compared to PLA as only the AND array is programmable. It is also easy to program a PAL compared to PLA as only AND must be programmed.

PAL devices consisted of a small PROM (programmable read-only memory) core and additional output logic used to implement particular desired logic functions with few components. Using specialized machines, PAL devices were "field-programmable". Each PAL device was "one-time programmable" (OTP), meaning that it could not be updated and reused after its initial programming. (MMI also offered a similar family called HAL, or "hard array logic", which were like PAL devices except that they were mask-programmed at the factory.)

The figure below shows a segment of an unprogrammed PAL. The input buffer with non-inverted and inverted outputs is used since each PAL must drive many AND Gates inputs. When the PAL is programmed, the fusible links (F1, F2, F3...F8) are selectively blown to leave the desired connections to the AND Gate inputs. Connections to the AND Gate inputs in a PAL are represented by **Xs**, as shown in Figure 6.4.

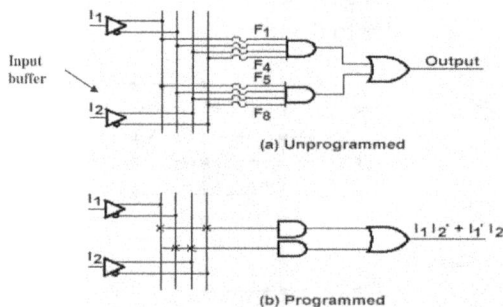

Figure 6.4: Programmable Array Logic.

Typical combinational PAL have 10 to 20 inputs and from 2 to 10 outputs with 2 to 8 AND gates driving each OR gate. PALs are also available which contain D flip-flops with inputs driven from the programming array logic. Such PAL provides a convenient way of realizing sequential networks. Figure 6.5 shows a segment of a sequential PAL.

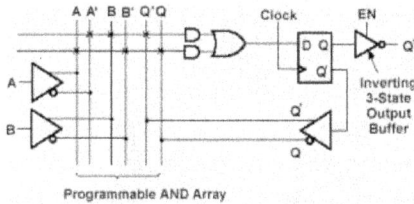

Figure 6.5: Sequential PAL.

The D flip-flop is driven from the OR gate, which is fed by two AND gates. The flip-flop output is fed back to the programmable AND array through a buffer. Thus the AND gate inputs can be connected to A, A′, B, B′, Q, or Q′. The Xs on the diagram shows the realization of the next-state equation.

$$Q+ = D = A'BQ' + AB'Q$$

The flip-flop output is connected to an inverting tri-state buffer, which is enabled when EN = 1

The Figures 6.6a&b shows a logic diagram for a typical sequential PAL.

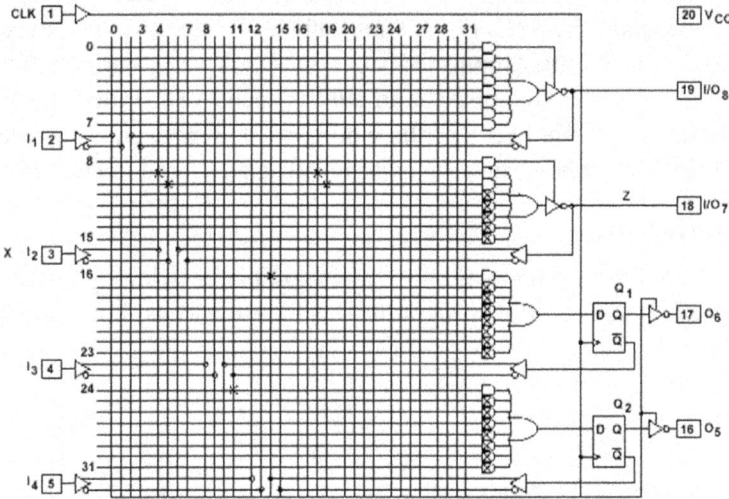

(a)

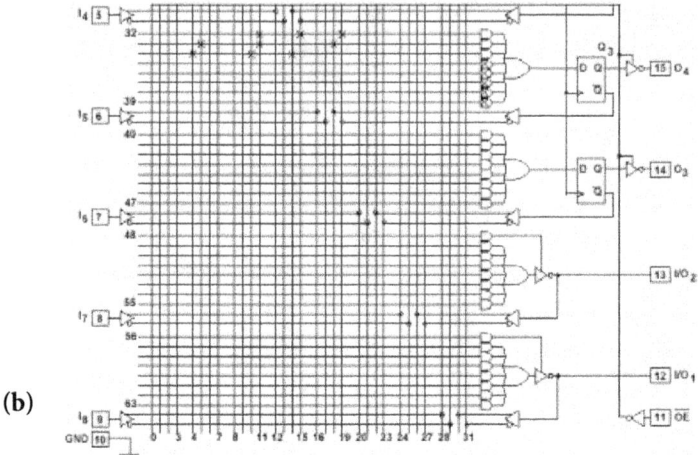

(b)

Figures 6.6a&b: Logic Diagram for a Typical Sequential PAL.

This PAL has an AND gate array with 16 input variables, and it has 4 D flip-flops. Each flip-flop output goes through a tri-state-inverting buffer (output pins 14-17). One input (pin 11) is used to enable these buffers. The rising edge of a common clock (pin 1) causes the flip-flops to change the state. Each D flip-flop input is driven from an OR gate, and each OR gate is fed from 8 AND gates.

The AND gate inputs can come from the external PAL inputs (pins2-9) or from the flip-flop outputs, which are fed back internally. In addition, there are four input/output (i/o) terminals (pins 12, 13,18 and 19), which can be used as either network outputs or as inputs to the AND gates. Thus each AND gate can have a maximum of 16 inputs (8 external inputs, 4 inputs fed back from the flip-flop outputs, and 4 inputs from the i/o terminals). When used as an output, each I/O terminal is driven from an inverting tri-state buffer. Each of these buffers is fed from an OR gate and each OR gate is fed from 7 AND gates. An eighth AND gate is used to enable the buffer.

PAL Architecture

The PAL architecture consists of two main components: logic plane and output logic microcells. The simplified programmable logic device is shown in Figure 6.7.

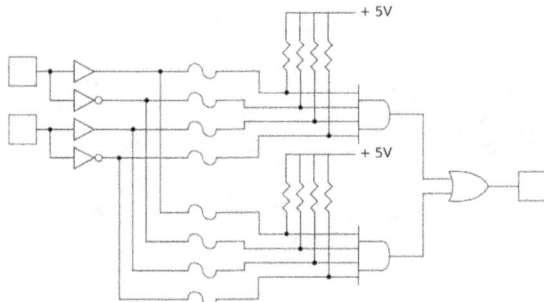

Figure 6.7: Simplified Programmable Logic Device.

Programmable Logic Plane

The programmable logic plane is a programmable (PROM) array that allows the signals present on the devices pins (or the logical complements of those signals) to be routed to an output logic macrocell. PAL devices have arrays of transistor cells arranged in a "fixed-OR, programmable-AND" plane used to implement "sum-of-products" binary logic equations for each of the outputs in terms of the inputs and either synchronous or asynchronous feedback from the outputs.

Output Logic

The early 20-pin PALs had 10 inputs and 8 outputs. The outputs were active low and could be registered or combinational. Members of the PAL family were available with various output structures called "output logic macrocells" or OLMCs. Prior to the introduction of the "V" (for "variable") series, the types of OLMCs available in each PAL were fixed at the time of manufacture. (The PAL16L8 had 8 combinational outputs and the PAL16R8 had 8 registered outputs. The PAL16R6 had 6 registered and 2 combinational while the PAL16R4 had 4 of each.) Each output could have up to 8 product terms (effectively AND gates), however, the combinational outputs used one of the terms to control a bidirectional output buffer. There were other combinations that had fewer outputs with more product term per output and were available with active high outputs. The 16X8 family or registered devices had an XOR gate before the register. There were also similar 24-pin versions of these PALs.

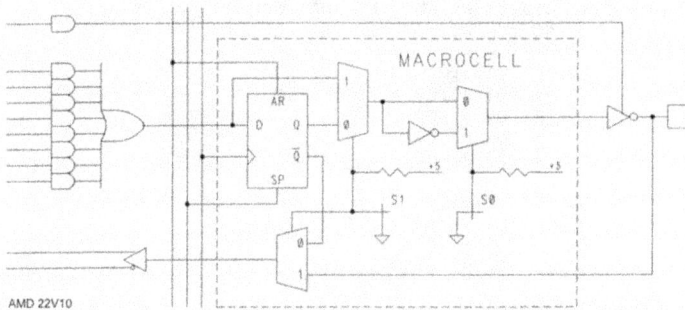

AMD 22V10

This fixed output structure often frustrated designers attempting to optimize the utility of PAL devices because output structures of different types were often required by their applications. (For example, one could not get 5 registered outputs with 3 active high combinational outputs.) So, in June 1983 AMD introduced the 22V10, a 24 pin device with 10 output logic macrocells. Each macrocell could be configured by the user to be combinational or registered, active high or active low. The number of product term allocated to an output varied from 8 to 16. This one device could replace all of the 24 pin fixed function PAL devices. Members of the PAL "V" ("variable") series included the PAL16V8, PAL20V8 and PAL22V10.

Programming PALs

PALs were programmed electrically using binary patterns and a special electronic programming system available from either the manufacturer or a third-party, such as DATA/IO. In addition to single-unit device programmers, device feeders and gang programmers were often used when more than just a few PALs needed to be programmed. (For large volumes, electrical programming costs could be eliminated by having the manufacturer fabricate a custom metal mask used to program the customers' patterns at the time of manufacture; MMI used the term "hard array logic" (HAL) to refer to devices programmed in this way.)

Programming Languages

Though some engineers programmed PAL devices by manually editing files containing the binary fuse pattern data, most opted to design their logic using a hardware description language(HDL) such as Data I/O's ABEL, Logical Devices' CUPL, or MMI's PALASM. These were computer-assisted design (CAD) (now referred to as "design automation") programs which translated (or "compiled") the designers' logic equations into binary fuse map files used to program (and often test) each device.

```
PAL16R4 PAL                 PAL DESIGN SPECIFICATION
CNT4SC
4 bit counter with synchronous clear
Michael Holley and Dave Pellerin
Clk  Clear  NC  NC  NC  NC  NC  NC  NC  GND
OE   NC     NC /Q3 /Q2 /Q1 /Q0  NC  NC  VCC

    Q3 := Clear
        + /Q3 * /Q2 * /Q1 * /Q0
        +  Q3 *  Q0
        +  Q3 *  Q1
        +  Q3 *  Q2

    Q2 := Clear
        + /Q2 * /Q1 * /Q0
        +  Q2 *  Q0
        +  Q2 *  Q1

    Q1 := Clear
        + /Q1 * /Q0
        +  Q1 *  Q0

    Q0 := Clear
        + /Q0

FUNCTION TABLE
OE Clear Clk   /Q0 /Q1 /Q2 /Q3
-------------------------------------
  L   H   C     L   L   L   L
  L   L   C     H   L   L   L
  L   L   C     L   H   L   L
  L   L   C     H   H   L   L
  L   L   C     L   L   H   L
  L   H   C     L   L   L   L
-------------------------------------
```

Combinational Logic Circuit

Combinational Circuits

Combinational logic is probably the easiest circuitry to design. The outputs from a combinational logic circuit depend only on the current inputs. The circuit has no remembrance of what it did at any time in the past.

Much of logic design involves connecting simple, easily understood circuits to construct a larger circuit that performs a much more complicated function. Several simple, often-used combinational logic circuits are the following:

Analysis Procedure

The bulk of the Combinational Analysis module is accessed through a single window of that name allowing you to view truth tables and Boolean expressions. This window can be opened in two ways.

Select Combinational Analysis and the current Combinational Analysis window will appear. If you haven't viewed the window before, the opened window will represent no circuit at all.

Only one Combinational Analysis window exists within Logisim, no matter how many projects are open. There is no way to have two different analysis windows open at once.

From a window for editing circuits, you can also request that Logisim analyze the current circuit by selecting the Analyze Circuit option from the Project menu. Before Logisim opens the window, it will compute Boolean expressions and a truth table corresponding to the circuit and place them there for you to view.

For the analysis to be successful, each input must be attached to an input pin, and each output must be attached to an output pin. Logisim will only analyze circuits with at most eight of each type, and all should be single-bit pins. Otherwise, you will see an error message and the window will not open.

In constructing Boolean expressions corresponding to a circuit, Logisim will first attempt to construct a Boolean expressions corresponding exactly to the gates in the circuit. But if the circuit uses some non-gate components (such as a multiplexer), or if the circuit is more than 100 levels deep (unlikely), then it will pop up a dialog box telling you those deriving Boolean expressions was impossible, and Logical will instead derive the expressions based on the truth table, which will be derived by quietly trying each combination of inputs and reading the resulting outputs.

After analyzing a circuit, there is no continuing relationship between the circuit and the Combinational Analysis window. That is, changes to the circuit will not be reflected in the window, nor will changes to the Boolean expressions and/or truth table in the window be reflected in the circuit. Of course, you are always free to analyze a circuit again; and, as we will see later, you can replace the circuit with a circuit corresponding to what appears in the Combinational Analysis window.

Limitations

Logical will not attempt to detect sequential circuits: If you tell it to analyze a sequential circuit, it will still create a truth table and corresponding Boolean expressions, although these will not accurately summarize the circuit behavior. (In fact, detecting sequential circuits is *probably impossible*, as it would amount to solving the Halting Problem. Of course, you might hope that Logical would make at least some attempt - perhaps look for flip-flops or cycles in the wires - but it does not.) As a result, the Combinational Analysis system should not be used indiscriminately: Only use it when you are indeed sure that the circuit you are analyzing is indeed combinational!

Logical will make a change to the original circuit that is perhaps unexpected: The Combinational Analysis system requires that each input and output have a unique name that conforms to the rules for Java identifiers. (Roughly, each character must either a letter or a digit, and the first character must be a letter. No spaces allowed!)

It attempts to use the pins' existing labels and to use a list of defaults if no label exists. If an existing label doesn't follow the Java-identifier rule, then Logical will attempt to extract a valid name from the label if at all possible. Incidentally, the ordering of the inputs in the truth table will match their top-down ordering in the original circuit, with ties being broken in left-right order.

Design Procedure

Logic circuits for digital systems may be combinational or sequential. A combinational circuit consists of logic gates whose outputs at any time are determined by combining the values of the applied inputs using logic operations. A combinational circuit performs an operation that can be specified logically by a set of the Boolean expression. In addition to using logic gates, sequential circuits employ elements that store bit values. Sequential circuit outputs are a function of inputs and the bit value in storage elements. These values, in turn, are a function of previously applied inputs and stored values. As a consequence, the outputs of a sequential circuit depend not only on the presently applied values of the inputs, but also on pas inputs, and the behavior of the circuit must be specified by a sequence in time of inputs and internally stored bit values. A combinational circuit consists of input variables, output variables, logic gates and interconnections. The interconnected logic gates accept signals from the inputs and generate signals at the output. The n input variables come from the environment of the circuit, and the m output variables are available for use by the environment. Each input and output variable exists physically as a binary signal that represents logic 1 or logic 0.

For n input variables, there are 2^n possible binary input combinations. For each binary combination of the input variables, there is one possible binary value on each output. Thus, a combinational circuit can be specified by a truth table that lists the output values for each combination of the input variables. A combinational circuit can also be described by m Boolean function, one for each output variable. Each such function is expressed as a function of the n input variables.

Combinational Circuit Design

The design of combinational circuit starts from a specification of the problem and culminates in a logic diagram or set of Boolean equations from which the logic diagram can be obtained. The procedure involves the following steps:

1. From the specifications of the circuit, determine the required number of inputs and outputs, and assign a letter symbol to each.

2. Derive the truth table that defines the required relationship between inputs and outputs.

3. Obtain the simplified Boolean functions of each output as a function of the input variables.

4. Draw the logic diagram.

5. Verify the correctness of the design.

Binary Adder-Subtractor

Half Adder/Subtractor

Figure 7.1 shows the configuration for a Half Adder/Subtractor. The first logic gate which is a XOR allows the circuit to do the complement of the binary bit when we want to do a subtraction. In the case that is needed to implement an addition the XOR keep the number the same.

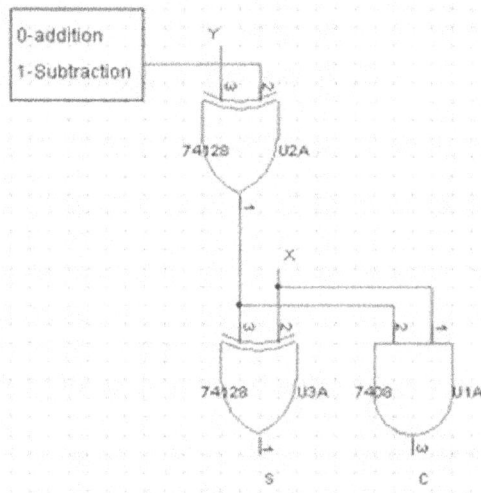

Figure 7.1: Half Adder/Subtractor.

Truth Table

A	B	C	S
0	0	0	0
0	1	0	1
1	0	0	1
1	1	1	0

Full Adder/Subtractor

This circuit adds two binary one-bit numbers. Also, it manages a carry that could come from another circuit (Figure 7.2).

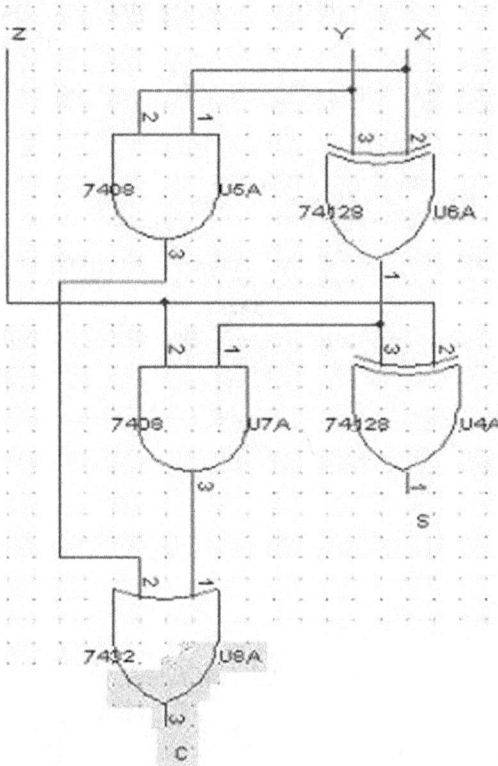

Figure 7.2: Full Adder Circuit.

$$[(AA)'*(BB)']' = A+B$$

X	Y	Z	C	S
0	0	0	0	0
0	0	1	0	1
0	1	0	0	1
0	1	1	1	0
1	0	0	0	1
1	0	1	1	0
1	1	0	1	0
1	1	1	1	1

This Figure 7.3 shows the configuration for a Full Adder/Subtractor. The box with the name of "Full Adder" has all the logic to do the addition and subtraction depending on the inputs that it receives. The XOR gate that is outside the box allows the circuit to do the complement of the binary bit when we want to do a subtraction. In the case that is needed to implement an addition, the XOR keep the number the same. In the previous graph, the internal logic for the box "Full Adder" is shown.

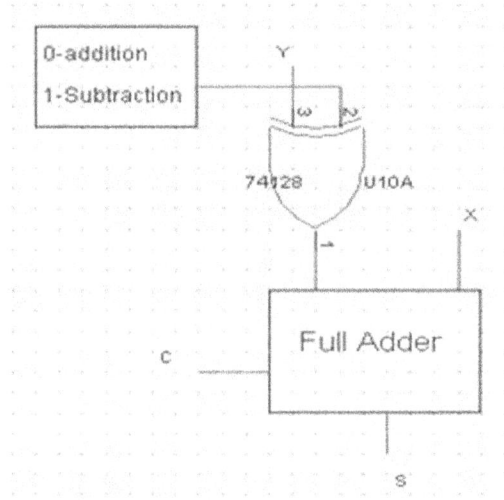

Figure 7.3: Full Adder.

3-bit Adder/Subtractor

The 3-bit adder/Subtractor was implemented with three Full adder circuits and three XOR gates outside which implemented the operation (addition/subtraction) selected by the user (Figure 7.4). In this circuit, we have three inputs for the first three bits binary number and three inputs for the second three bits binary number. When the addition is selected the XOR gates keep the binary numbers the same and add them together. On the other hand, when a subtraction is selected the XOR gates complement the second number which is represented with a "Y" and after that, the Full Adder circuit adds both numbers together. The 3-bit Adder/Subtractor circuit has four outputs. The first three outputs represent the three bits that were the result of the subtraction or addition. The last bit represents a carry.

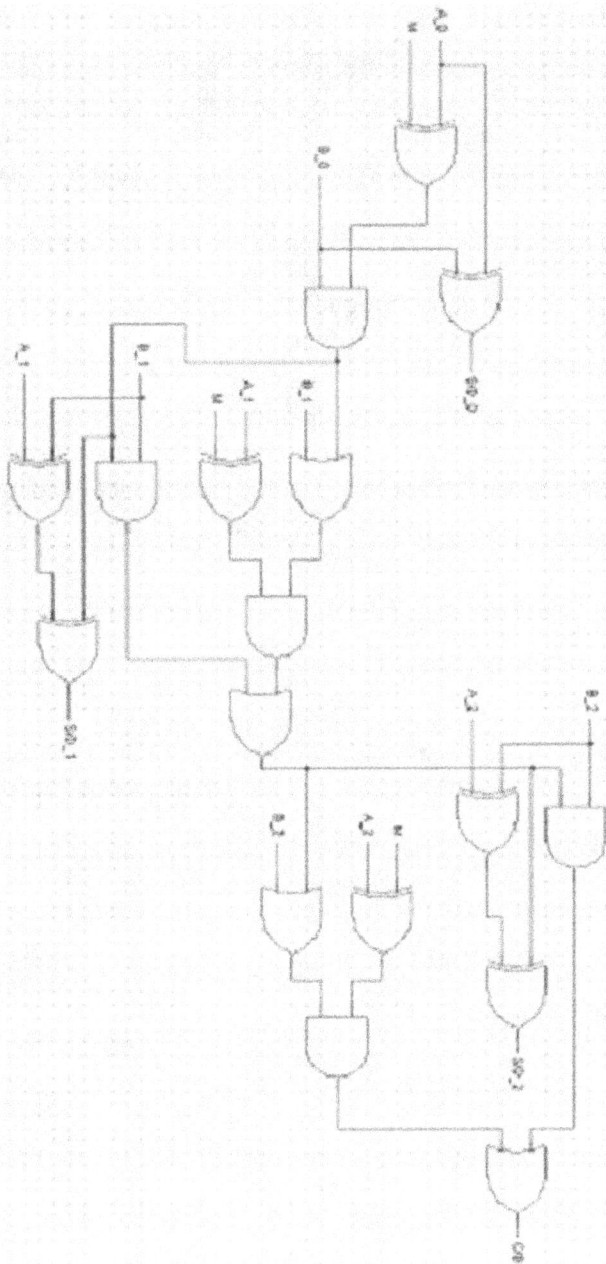

Figure 7.4: 3-bit Adder/Subtractor.

Binary Multiplier

Binary multiplication uses the same technique as decimal multiplication. In fact, binary multiplication is much easier because each digit we multiply by is either zero or one. Consider the simple problem of multiplying 110_2 by 10_2. We can use this problem to review some terminology and illustrate the rules for binary multiplication.

1. First, we note that 110_2 is our multiplicand and 10_2 is our multiplier.

   110
   × 10

2. We begin by multiplying 110_2 by the rightmost digit of our multiplier which is 0. Any number times zero is zero, so we just write zeros below.

   110
   × 10
   000

3. Now we multiply the multiplicand by the next digit of our multiplier which is 1. To perform this multiplication, we just need to copy the multiplicand and shift it one column to the left as we do in decimal multiplication.

   110
   × 10
   000
   110

4. Now we add our results together. The product of our multiplication is 1100_2.

   110
   × 10
   000
   110
   1100

When performing binary multiplication, remember the following rules:

1. Copy the multiplicand when the multiplier digit is 1. Otherwise, write a row of zeros.

2. Shift your results one column to the left as you move to a new multiplier digit.

3. Add the results together using binary addition to find the product.

Magnitude Comparator

74L85 4-bit magnitude comparator is given in Figure 7.5.

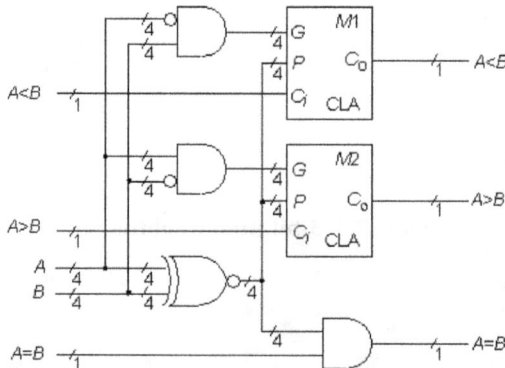

Figure 7.5: 74L85 4-Bit Magnitude Comparator.

11 inputs; 3 outputs; 33 gates;

The 74L85 magnitude comparator can be functionally modeled as above. This is a simplification of implementing a magnitude comparator by a carry function with an inverted input bus as shown.

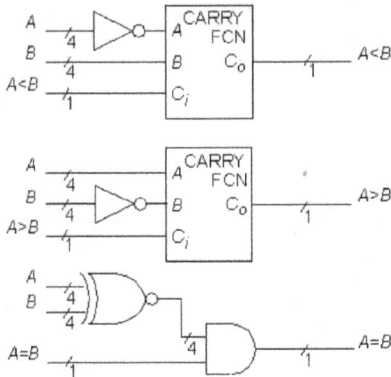

Using this concept, common elements of the three comparator functions $A < B$, $A > B$ and $A = B$ are combined to construct the model shown above, which maps directly onto the gate-level realization of the 74L85.

Decoders

Introduction

In both the multiplexer and the demultiplexer, part of the circuits *decode* the address inputs, *i.e.* it translates a binary number of n digits to 2^n outputs, one of which (the one that corresponds to the value of the binary number) is 1 and the others of which are 0.

It is sometimes advantageous to separate this function from the rest of the circuit, since it is useful in many other applications. Thus, we obtain a new combinatorial circuit that we call the *decoder*. It has the following truth table (for $n = 3$):

a2	a1	a0	d7	d6	d5	d4	d3	d2	d1	d0
0	0	0	0	0	0	0	0	0	0	1
0	0	1	0	0	0	0	0	0	1	0
0	1	0	0	0	0	0	0	1	0	0
0	1	1	0	0	0	0	1	0	0	0
1	0	0	0	0	0	1	0	0	0	0
1	0	1	0	0	1	0	0	0	0	0
1	1	0	0	1	0	0	0	0	0	0
1	1	1	1	0	0	0	0	0	0	0

Here is the circuit Figure 7.6 for the decoder:

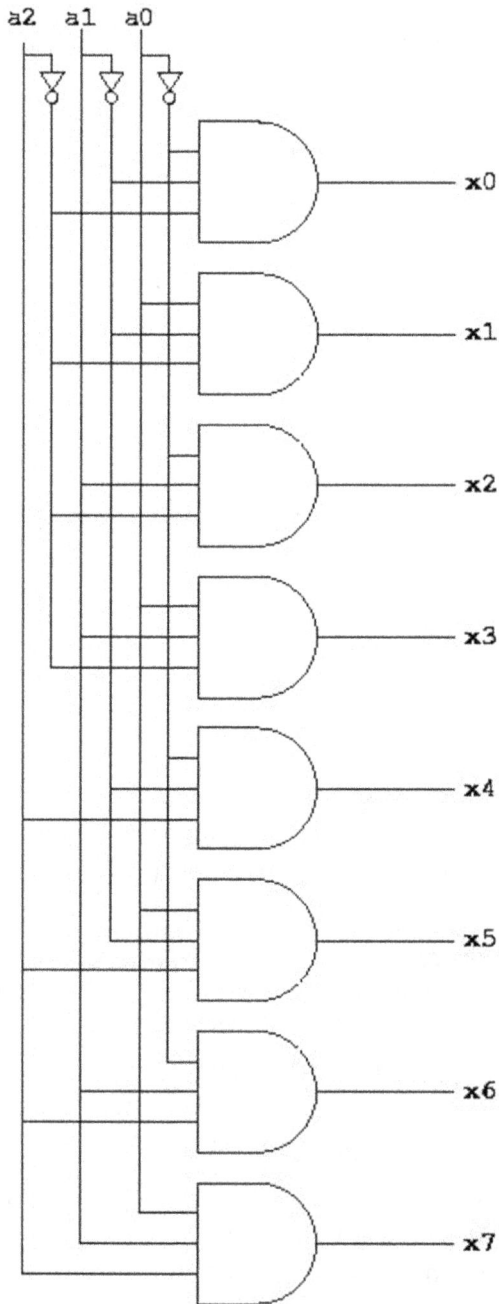

Figure 7.6: Decoder Circuit.

Decoder Generation

X86's decoder is generated using an ISA description like all the other ISAs, although how it does that is a bit different. Most of the instructions for most of the other ISAs are defined by passing chunks of code that perform the instruction into an instruction format. The format is basically a template which puts wraps that bit of code in the structure needed to support it and you have your instruction object. Because almost all of the instructions in x86 are microcoded and many can be encoded in multiple ways and hence appear in the decoder more than once, and because the same non-trivial decoding rules apply to many different instructions, X86 uses the decoder as a layer of indirection and defines the majority of its instructions elsewhere (Figure 7.7).

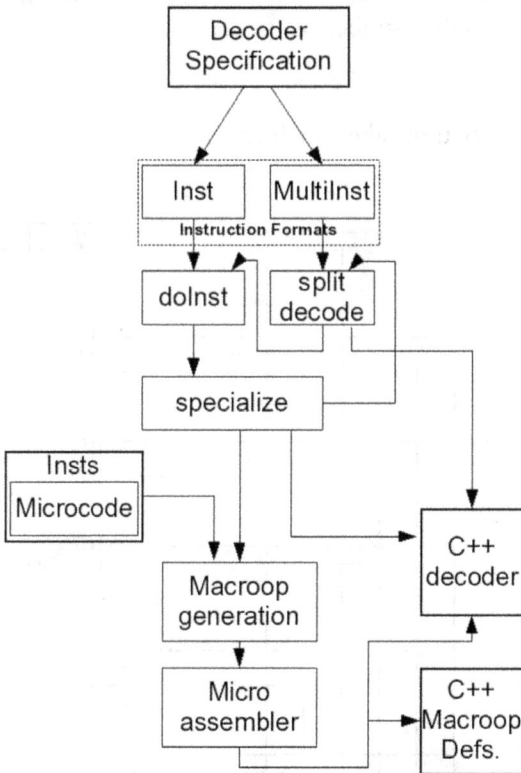

Figure 7.7: X86's Decoder.

X86 almost exclusively uses only two different instruction formats, Inst and MultiInst. MultiInst is just a compact way of describing multiple related Insts. An Inst essentially selects an instruction like XOR and provides a specification for its operands. Inside the instruction format, the instruction name and operand specification are passed to a python function called "specializeInst" which figures out what to do with it. If the operand specification describes more than one version of the instruction, for instance, one that uses memory and one that uses registers, the instruction's information are passed into another function, "doSplitDecode", which separates out those versions and passes each individually back through the same system. This goes on until the instructions have been fully split out and the code has been generated to figure out what version to use. As a nice bonus, the MultiInst format doesn't add much complexity to this model since it can simply jump right into doSplitDecode and continue as normal. There is one additional format for string instructions that works similar to MultiInst, except instead of specializing the instruction based on its operands, it specializes it based on its prefixes.

At this point, the code for selecting the right version of an instruction is put into the C++ decoder function. Almost all of this function is built this way, with the minor exception of small bits of logic that glue everything together and make large- scale distinctions the number of opcode bytes.

3 to 8 Decoder

The 3 to 8 decoder unit takes 3 address lines as input and outputs 8 address enable lines. (Figure 7.8).

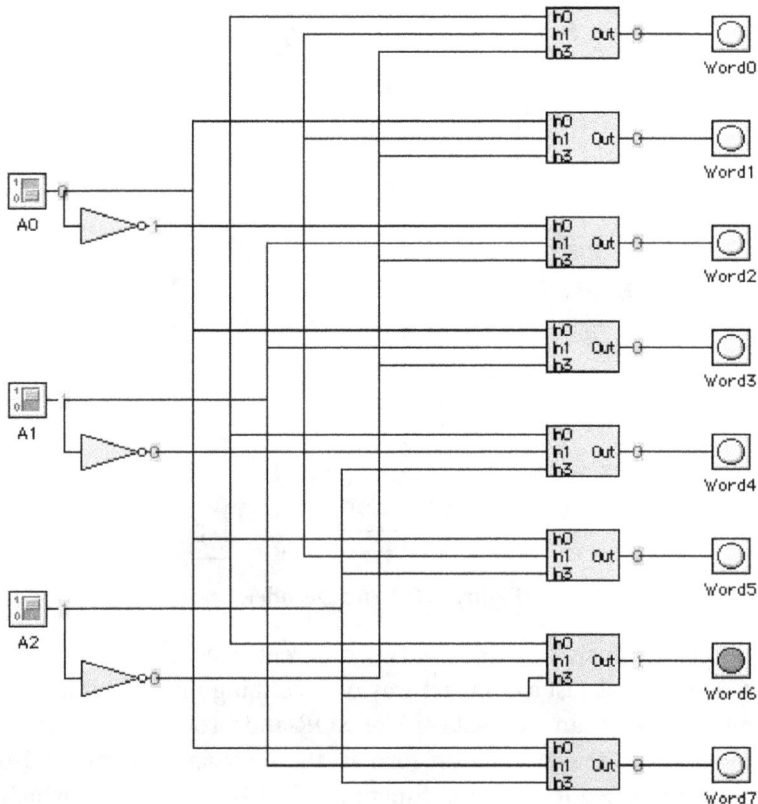

Figure 7.8: 3: 8 Decoder.

Which of the 8 output is enabled is dependent upon the configuration of the 3 address line. If A0, A1, and A2 are all 0, then address 0 is enabled. If A0 = 0, A1 = 1, A2 = 0, then address 3 is enables. With 3 address lines, the number of words that can be addressed is 8 ($2^3=8$)

Encoders

An encoder is a circuit that changes a set of signals into a code. Let's begin making a 2-to-1 line encoder truth table by reversing the 1-to-2 decoder truth table.

D_1	D_0	A
0	1	0
1	0	1

This truth table is a little short. A complete truth table would be

D_1	D_0	A
0	0	
0	1	0
1	0	1
1	1	

Truth Table

One question we need to answer is what to do with those other inputs? Do we ignore them? Do we have them generate an additional error output? In many circuits, this problem is solved by adding sequential logic in order to know not just what input is active but also which order the inputs became active. A more useful application of combinational encoder design is a binary to 7-segment encoder. The seven segments are given according:

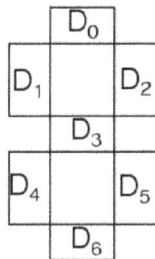

Our truth table is:

I_3	I_2	I_1	I_0	D_6	D_5	D_4	D_3	D_2	D_1	D_0
0	0	0	0	1	1	1	0	1	1	1
0	0	0	1	0	0	1	0	0	1	0
0	0	1	0	1	0	1	1	1	0	1
0	0	1	1	1	0	1	1	0	1	1
0	1	0	0	0	1	1	1	0	1	0
0	1	0	1	1	1	0	1	0	1	1
0	1	1	0	1	1	0	1	1	1	1
0	1	1	1	1	0	1	0	0	1	0
1	0	0	0	1	1	1	1	1	1	1
1	0	0	1	1	1	1	1	0	1	1

Truth Table

Deciding what to do with the remaining six entries of the truth table is easier with this circuit. This circuit should not be expected to encode an undefined combination of inputs, so we can leave them as "don't care" when we design the circuit. The Boolean equations

$$D_0 = I_3 + I_1 + \bar{I_3}\bar{I_2}I_1\bar{I_0} + \bar{I_3}I_2\bar{I_1}I_0$$
$$D_1 = I_3 + \bar{I_2}I_1 + I_2\bar{I_1} + I_2\bar{I_0}$$
$$D_2 = I_2 + \bar{I_3}I_2\bar{I_1}\bar{I_0} + \bar{I_3}I_2I_1\bar{I_0}$$
$$D_3 = I_3 + I_1\bar{I_0} + I_2\bar{I_1}$$
$$D_4 = I_1\bar{I_0} + \bar{I_2}I_1\bar{I_0}$$
$$D_5 = I_3 + I_2 + I_0$$
$$D_6 = I_3 + I_1\bar{I_0} + \bar{I_3}I_2I_1 + \bar{I_3}I_2\bar{I_1}\bar{I_0} + \bar{I_3}\bar{I_2}I_1\bar{I_0}$$

and the circuit is

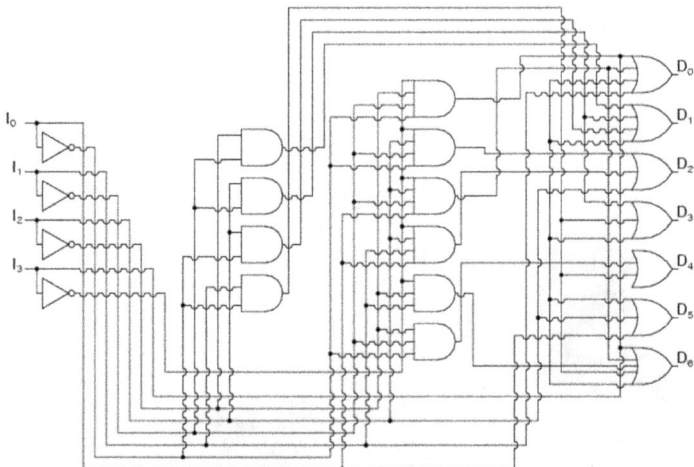

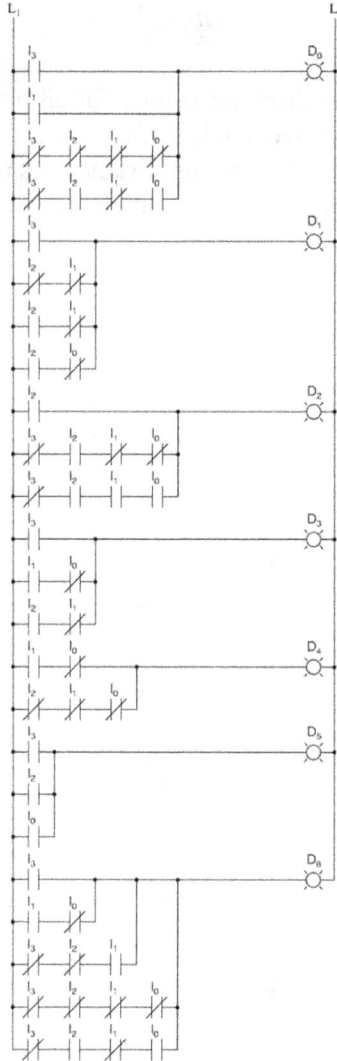

Multiplexes

Introduction

A multiplexer is a combinatorial circuit that is given a certain number (usually a power of two) *data inputs*, let us say 2^n, and n *address inputs* used as a binary number to select one of the data inputs. The multiplexer has a single output, which has the same value as the selected data input.

In other words, the multiplexer works like the input selector of a home music system. Only one input is selected at a time, and the selected input is transmitted to the single output. While on the music system, the selection of the input is made manually, the multiplexer chooses its input based on a binary number, the address input.

The truth table for a multiplexer is huge for all but the smallest values of n. We therefore use an abbreviated version of the truth table in which some inputs are replaced by '-' to indicate that the input value does not matter. Here is such an abbreviated truth table for n = 3. The full truth table would have $2^{(3+23)} = 2048$ rows.

a_2	a_1	a_0	d_7	d_6	d_5	d_4	d_3	d_2	d_1	d_0	x
-	-	-	-	-	-	-	-	-	-	-	-
0	0	0	-	-	-	-	-	-	-	0	0
0	0	0	-	-	-	-	-	-	-	1	1
0	0	1	-	-	-	-	-	-	0	-	0
0	0	1	-	-	-	-	-	-	1	-	1
0	1	0	-	-	-	-	-	0	-	-	0
0	1	0	-	-	-	-	-	1	-	-	1
0	1	1	-	-	-	-	0	-	-	-	0
0	1	1	-	-	-	-	1	-	-	-	1
1	0	0	-	-	-	0	-	-	-	-	0
1	0	0	-	-	-	1	-	-	-	-	1
1	0	1	-	-	0	-	-	-	-	-	0
1	0	1	-	-	1	-	-	-	-	-	1
1	1	0	-	0	-	-	-	-	-	-	0
1	1	0	-	1	-	-	-	-	-	-	1
1	1	1	0	-	-	-	-	-	-	-	0
1	1	1	1	-	-	-	-	-	-	-	1

We can abbreviate this table even more by using a letter to indicate the value of the selected input, like this:

a_2	a_1	a_0	d_7	d_6	d_5	d_4	d_3	d_2	d_1	d_0	x
-	-	-	-	-	-	-	-	-	-	-	-
0	0	0	-	-	-	-	-	-	-	c	c
0	0	1	-	-	-	-	-	-	c	-	c
0	1	0	-	-	-	-	-	c	-	-	c
0	1	1	-	-	-	-	c	-	-	-	c
1	0	0	-	-	-	c	-	-	-	-	c
1	0	1	-	-	c	-	-	-	-	-	c
1	1	0	-	c	-	-	-	-	-	-	c
1	1	1	c	-	-	-	-	-	-	-	c

The same way we can simplify the truth table for the multiplexer, we can also simplify the corresponding circuit. Indeed, our simple design method would yield a very large circuit. The simplified circuit looks like this (Figure 7.9):

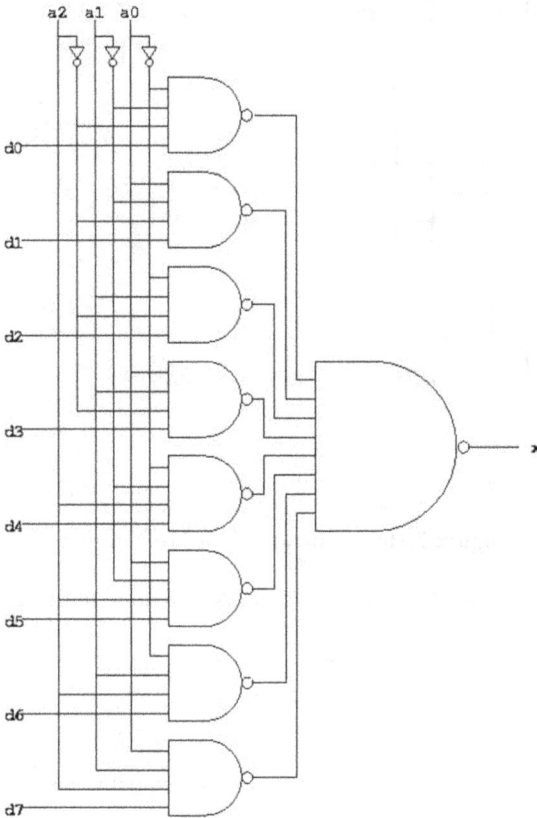

Figure 7.9: Multiplexer.

A multiplexer performs the function of selecting the input on any one of 'n' input lines and feeding this input to one output line.

Multiplexers are used as one method of reducing the number of integrated circuit packages required by a particular circuit design. This in turn, reduces the cost of the system.

Assume that we have four lines, **C0, C1, C2 and C3**, which are to be multiplexed on a single line, **Output (f)**. The four input lines are also known as the **Data Inputs**. Since there are four inputs, we will need two additional inputs to the multiplexer, known as the **Select Inputs**, to select which of the **C** inputs is to appear at the output. Call these select lines **A and B**. The gate implementation of a 4-line to 1-line multiplexer is shown in Figure 7.10.

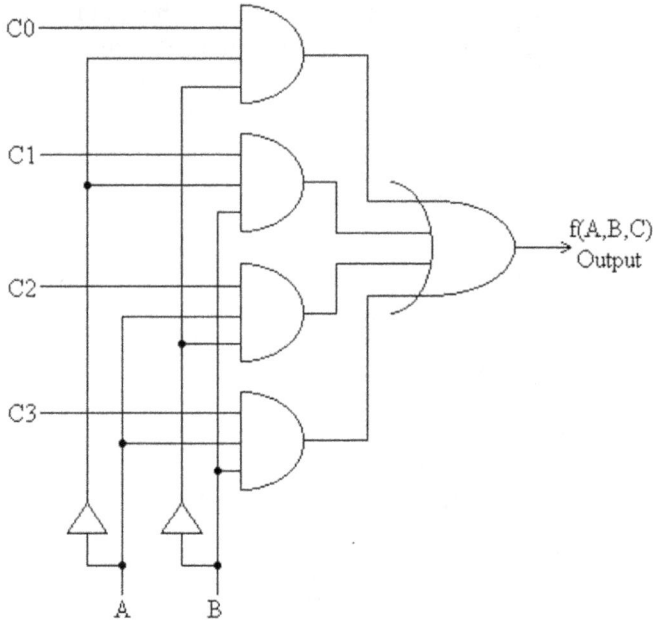

Figure 7.10: 4-Line to 1-Line Multiplexer.

The circuit symbol for the above multiplexer is depicted in Figure 7.11.

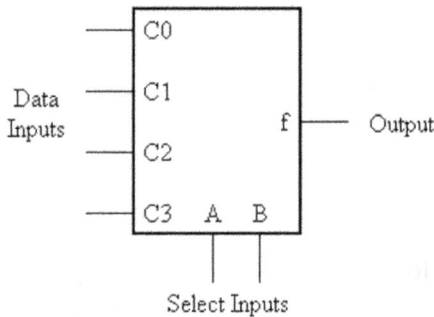

Figure 7.11: 4-Line to 1-Line Multiplexer Symbol.

4 Inputs Multiplexer

The multiplexer concept is not limited to two data inputs. If we add a second addressing input, B, we can control as many as four data inputs, as shown to the left. A third and fourth addressing input will allow the multiplexer to control eight or sixteen inputs, respectively (Figure 7.12).

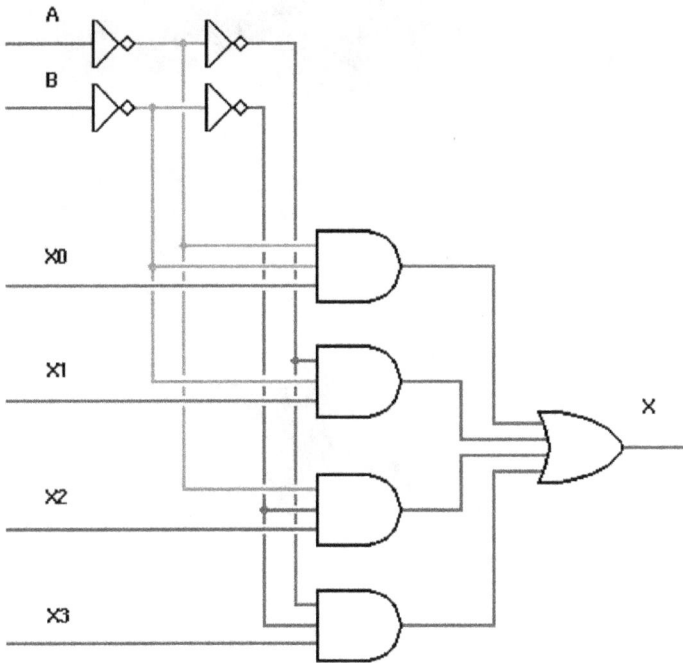

Figure 7.12: 4-Input Multiplexer.

Inputs A and B are the addressing inputs to this multiplexer. They select which of the four data inputs will be transmitted to the final output, X. If the data inputs are to be multiplexed for transmission to a distant location, the inputs must cycle through all four possible addresses more than twice for every single cycle of each of the data inputs. Otherwise, the input data cannot be reconstructed accurately at the receiving end.

Unit 8

Synchronous Sequential Logic

Sequential Circuits

Digital electronics is classified into combinational logic and sequential logic. Combinational logic output depends on the inputs levels, whereas sequential logic output depends on stored levels and also the input levels (Figure 8.1).

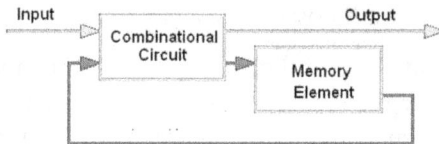

Figure 8.1: Combinational Logic Circuit.

The Figure 8.2 shows a way to consider sequential circuits.

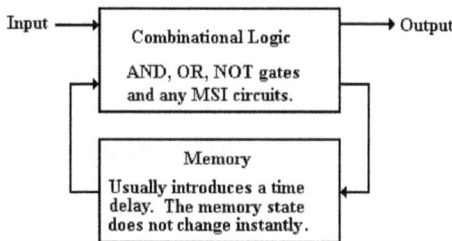

Figure 8.2: Sequential Logic Includes Combinational Logic and Memory.

Sequential circuits can be characterized into two broad classes – synchronous and asynchronous. As a general rule, asynchronous circuits are faster but much harder to design. We shall focus totally on synchronous circuits.

By Q(T) we denote the state of a sequential circuit at time T – this is basically its memory. We watch the state of the circuit change from Q(T) to Q(T + 1) as the clock ticks. The constraint on synchronous circuits is that the state of the circuit changes after the input, thus we have a typical sequence as follows:

1. At time T, we have input (denoted by X) and state Q(T).

2. As a result of the input X and state Q(T), a new state is computed, This becomes available to the input only at the time (T + 1) and so is called Q(T + 1).

The fact that the new state, computed as a result of X and Q(T), is not available to the input of the sequential circuit until the next time step greatly facilitates the design and analysis of the specific circuit.

Circuit Analysis

Circuit analysis begins with a circuit diagram or a black box and ends with an identification of the sequential circuit implemented by the device – normally a truth table. The steps are:

1. Identify the inputs and the outputs

2. Express each output as a Boolean function of the inputs and the present state Q(T)

3. Identify the circuit if possible.

We first study the analysis of digital circuits, and then we study their design. There are a number of steps in the analysis of a circuit. Where to begin depends on what one has. When given a circuit diagram, the following steps are used to begin the analysis.

1. Determine the inputs and outputs of the circuit. Assign variables to represent these.

2. Determine the inputs and outputs of the flip-flops.

3. Construct the Next State and Output Tables.

4. Construct the State Diagram.

5. If possible, identify the circuit. There are no good rules for this step.

Consider the circuit given in Figure 8.3. We want to discover what it does.

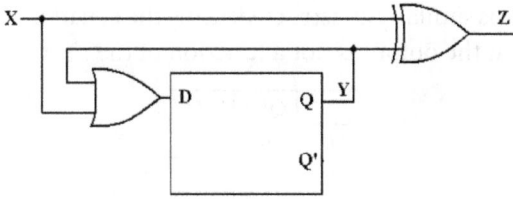

Figure 8.3: Circuit to be Analyzed

Steps of Analysis

Step 1: Identify and Label the Inputs, Outputs, and the Internal States

We use the following variables in the analysis of this circuit with a single flip-flop. X denotes the input, Y denotes the output of the flip-flop (Y' also), and Z the output of the circuit. If we had more than one flip-flop, we would label the flip-flop with a number beginning at 0 and use that as a subscript, so flip-flop 0 would have output Y_0, etc.

Step 2: Determine the Inputs and Outputs of the Flip-flops

The next step is to determine the equations for Z, the output, and D, the input to the flip-flop. By inspection, we determine the following for the equations:

$Z = X \oplus Y$

$D = X + Y$

Step 3: Construct the Next State and Output Tables

We begin this state by recalling the characteristic table of each flip-flop that is used in the design. Here we have only one flip-flip, a D with a very simple characteristic table that is better represented as an equation: $Q(t+1) = D$ – the next state is what you put in now.

Noting that $Q(t) = Y$ (the state of a flip-flop is also its output) we construct the following Next-State diagram for the flip-flop, based on the characteristic table of a D flip-flop and the equation we derived for the D input: $D = X + Y$.

One simple caution here is that the input to a flip-flop is a function of the present state only, having nothing to do with the next state (as we have no crystal balls). Thus $Y = Q(t)$. Here is the present state (PS) / next state (NS) diagram for the circuit.

X	Q(t) = Y	D = X + Y	Q(t+1)
0	0	0	0
0	1	1	1
1	0	1	1
1	1	1	1

The output table is similarly constructed, using the equation $Z = X \oplus Y = Z = X \oplus Q(t)$. Again, note that the output is not a function of the next state.

X	Q(t)	Z
0	0	0
0	1	1
1	0	1
1	1	0

These two tables are combined to form the transition / output table.

X	Q(t) = Y	D = X + Y	Q(t+1) / Z
0	0	0	0 / 0
0	1	1	1 / 1
1	0	1	1 / 1
1	1	1	1 / 0

At this point, we should produce a state table in the standard format. This involves assigning labels to each of the two states, currently identified only as $Q(t) = 0$ and $Q(t) = 1$. For lack of anything more imaginative, we label the states 0 and 1.

Present State	Next State/Output	
	X = 0	X = 1
0	0 / 0	1 / 1
1	1 / 1	1 / 0

Table: State Table for Circuit to be Analyzed

Step 4: Construct the State Diagram

The final step in the process may be the creation of the state diagram (Figure 8.4).

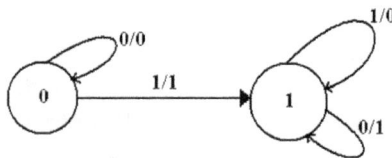

Figure 8.4: State Diagram.

At this point, we have a complete description of the circuit. It may be possible to proceed from this diagram to obtain an understanding of what the circuit does.

Step 5: If possible, identify the circuit

This circuit is a 2-state device, with memory represented by one bit. The circuit stays in state 0 from the start until a 1 is input at which time it transitions to state

1 and remains there. Note the relation of the output to the input, depending on the state of the machine.

Input	Q(t)	Output	
0	0	0	For Q(t) = 0, the output is X
1	0	1	
0	1	1	For Q(t) = 1, the output is X'.
1	1	0	

A verbal description of the circuit is then that it copies its input to the output until the first 1 is encountered in the input stream. After that event, all input is output as complemented. What this circuit does is take the two's-complement of a binary integer, presented to the circuit least-significant bit first. It is easy to prove that such a strategy produces the two's complement of a number. First, consider a number ending in 1, say $X_n X_{n-1} .... X_2 X_1 1$.

The one's complement of this number is $X_n' X_n - 1' .... X_2' X_1' 0$. Adding 1 to this produces the number $X_n' X_n - 1' .... X_2' X_1' 1$, in which the least significant 1 is copied and the remaining bits are complemented. The proof is completed by supposing that the number terminates in a one or more zeroes; *i.e.*, its least significant bits are 10 .... 0, where the count of zeroes is not important. The one's-complement of this number will end with 01 .... 1, where each zero in the original has turned to a 1. But 01 .... 1 + 1 = 10 .... 0, and up to the least significant 1 the two's-complement is a copy of the bits in the integer itself. Note that there is no carry out of the addition that produced the right-most 1, so the remainder of the integer is formed by the one's-complement. Thus we have a complete description of the circuit.

Latches

When showing either latches or flip–flops as circuit elements, it is undesirable to show the "internals" of the device. Here are the symbols for SR and D latches (Figure 8.5).

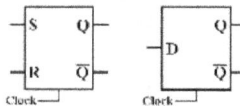

Figure 8.5: Symbols for SR and D Latches.

The S-R Latch

A bistable multivibrator has *two* stable states, as indicated by the prefix *bi* in its name. Typically, one state is referred to *asset* and the other as *reset*. The simplest bistable device, therefore, is known as a *set-reset*, or S-R, latch.

To create an S-R latch, we can wire two NOR gates in such a way that the output of one feeds back to the input of another, and visa-versa, like this (Figure 8.6):

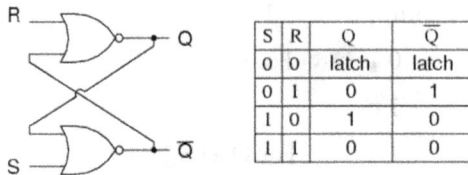

S	R	Q	$\overline{Q}$
0	0	latch	latch
0	1	0	1
1	0	1	0
1	1	0	0

Figure 8.6: S-R Latch.

The Q and not-Q outputs are supposed to be in opposite states. I say "supposed to" because making both the S and R inputs equal to 1 results in both Q and not-Q being 0. For this reason, having both S and R equal to 1 is called an *invalid* or *illegal* state for the S-R multivibrator. Otherwise, making S = 1 and R = 0 "sets" the multivibrator so that Q = 1 and not-Q = 0. Conversely, making R = 1 and S = 0 "resets" the multivibrator in the opposite state. When S and R are both equal to 0, the multivibrator's outputs "latch" in their prior states. Note how the same multivibrator function can be implemented in ladder logic, with the same results:

By definition, a condition of Q = 1 and not-Q = 0 is set. A condition of Q = 0 and not-Q = 1 is *reset*. These terms are universal in describing the output states of any multivibrator circuit (Figure 8.7).

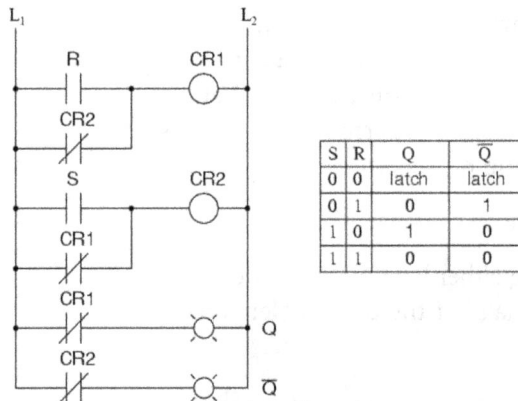

S	R	Q	$\overline{Q}$
0	0	latch	latch
0	1	0	1
1	0	1	0
1	1	0	0

Figure 8.7: Multivibrator Circuit.

The astute observer will note that the initial power-up condition of either the gate or ladder variety of S-R latch is such that both gates (coils) start in the de-energized mode. As such, one would expect that the circuit will start up in an invalid condition, with both Q and not-Q outputs being in the same state. Actually, this is true! However, the invalid condition is unstable with both S and R inputs inactive, and the circuit will quickly stabilize in either the set or reset condition because one gate (or relay) is bound to react a little faster than the other. If both gates (or coils) were *precisely identical*, they would oscillate between high and low like an astable multivibrator upon power-up without ever reaching a point of stability! Fortunately for cases like this, such a precise match of components is a rare possibility.

It must be noted that although an astable (continually oscillating) condition would be extremely rare, there will most likely be a cycle or two of oscillation in the above circuit, and the final state of the circuit (set or reset) after power-up would be unpredictable. The root of the problem is a *race condition* between the two relays CR_1 and CR_2.

A race condition occurs when two mutually-exclusive events are simultaneously initiated through different circuit elements by a single cause. In this case, the circuit elements are relays CR_1 and CR_2, and their de-energized states are mutually exclusive due to the normally-closed interlocking contacts. If one relay coil is de-energized, its normally-closed contact will keep the other coil energized, thus maintaining the circuit in one of two states (set or reset). Interlocking prevents *both* relays from latching. However, if *both* relay coils start in their de-energized states (such as after the whole circuit has been de-energized and is then powered up) both relays will "race" to become latched on as they receive power (the "single cause") through the normally-closed contact of the other relay. One of those relays will inevitably reach that condition before the other, thus opening its normally-closed interlocking contact and de-energizing the other relay coil. Which relay "wins" this race is dependent on the physical characteristics of the relays and not the circuit design, so the designer cannot ensure which state the circuit will fall into after power-up.

Race conditions should be avoided in circuit design primarily for the unpredictability that will be created. One way to avoid such a condition is to insert a time-delay relay into the circuit to disable one of the competing relays for a short time, giving the other one a clear advantage. In other words, by purposely slowing down the de-energization of one relay, we ensure that the other relay will always "win" and the race results will always be predictable. Here is an example of how a time-delay relay might be applied to the above circuit to avoid the race condition (Figure 8.8):

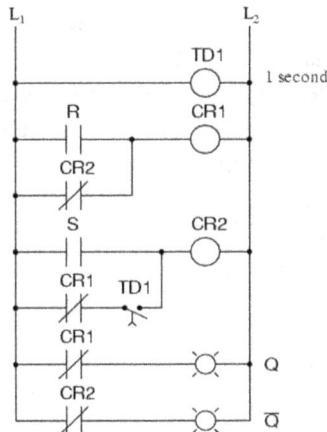

Figure 8.8: Time-Delay Relay

When the circuit powers up, time-delay relay contact TD$_1$ in the fifth rung down will delay closing for 1 second. Having that contact open for 1 second prevents relay CR$_2$ from energizing through contact CR$_1$ in its normally-closed state after power-up. Therefore, relay CR$_1$ will be allowed to energize first (with a 1-second head start), thus opening the normally-closed CR$_1$contact in the fifth rung, preventing CR$_2$ from being energized without the S input going active. The end result is that the circuit powers up cleanly and predictably in the reset state with S = 0 and R = 0.

It should be mentioned that race conditions are not restricted to relay circuits. Solid-state logic gate circuits may also suffer from the ill effects of race conditions if improperly designed. Complex computer programs, for that matter, may also incur race problems if improperly designed. Race problems are a possibility for any sequential system, and may not be discovered until sometime after initial testing of the system. They can be very difficult problems to detect and eliminate.

A practical application of an S-R latch circuit might be for starting and stopping a motor, using normally-open, momentary pushbutton switch contacts for both *start* (S) and *stop* (R) switches, then energizing a motor contactor with either a CR$_1$ or CR$_2$ contact (or using a contactor in place of CR$_1$ or CR$_2$). Normally, a much simpler ladder logic circuit is employed, such as this (Figure 8.9):

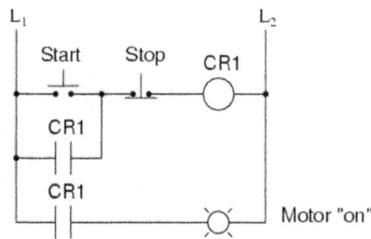

Figure 8.9: Ladder Logic Circuit.

In the above motor start/stop circuit, the CR$_1$ contact in parallel with\ the *start* switch contact is referred to as a "seal-in" contact, because of it "seals" or latches control relay CR$_1$ in the energized state after the *start* switch has been released. To break the "seal," or to "unlatch" or "reset" the circuit, the *stop* pushbutton is pressed, which de-energizes CR$_1$ and restores the seal-in contact to its normally open status. Notice, however, that this circuit performs much the same function as the S-R latch. Also, note that this circuit has no inherent instability problem (if even a remote possibility) as does the double relay S-R latch design.

In semiconductor form, S-R latches come in prepackaged units so that you don't have to build them from individual gates. They are symbolized as such (Figure 8.10):

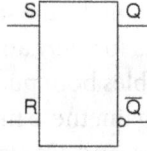

Figure 8.10: S-R Latch.

The Gated S-R Latch

It is sometimes useful in logic circuits to have a multivibrator which changes state only when certain conditions are met, regardless of its S and R input states. The conditional input is called the *enable*, and is symbolized by the letter E. Study the following example to see how this works (Figure 8.11):

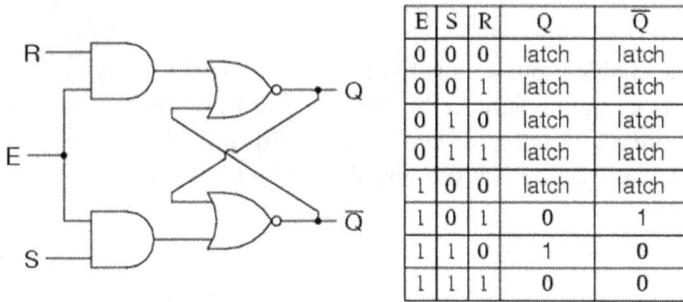

E	S	R	Q	$\bar{Q}$
0	0	0	latch	latch
0	0	1	latch	latch
0	1	0	latch	latch
0	1	1	latch	latch
1	0	0	latch	latch
1	0	1	0	1
1	1	0	1	0
1	1	1	0	0

Figure 8.11: S and R Input States.

When the E = 0, the outputs of the two AND gates are forced to 0, regardless of the states of either S or R. Consequently, the circuit behaves as though S and R were both 0, latching the Q and not-Q outputs in their last states. Only when the enable input is activated (1) will the latch respond to the S and R inputs. Note the identical function in ladder logic (Figure 8.12):

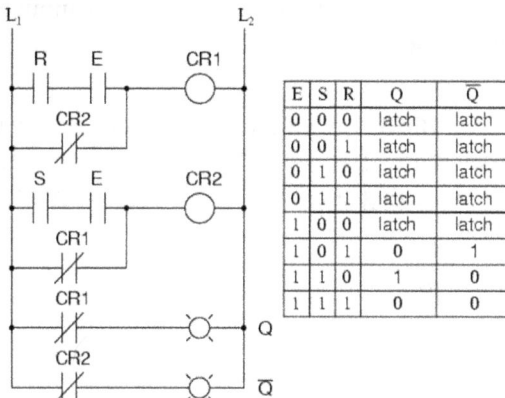

E	S	R	Q	$\bar{Q}$
0	0	0	latch	latch
0	0	1	latch	latch
0	1	0	latch	latch
0	1	1	latch	latch
1	0	0	latch	latch
1	0	1	0	1
1	1	0	1	0
1	1	1	0	0

Figure 8.12: S and R Ladder Logic.

A practical application of this might be the same motor control circuit (with two normally-open pushbutton switches for *start* and *stop*), except with the addition of a master lockout input (E) that disables both pushbuttons from having control over the motor when it's low (0). Once again, these multivibrator circuits are available as prepackaged semiconductor devices, and are symbolized as such:

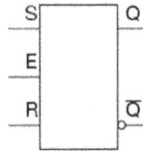

It is also common to see the enable input designated by the letters "EN" instead of just "E."

D Latch

Since the enable input on a gated S-R latch provides a way to latch the Q and not-Q outputs without regard to the status of S or R, we can eliminate one of those inputs to create a multivibrator latch circuit with no "illegal" input states. Such a circuit is called a D latch, and its internal logic looks like this (Figure 8.13):

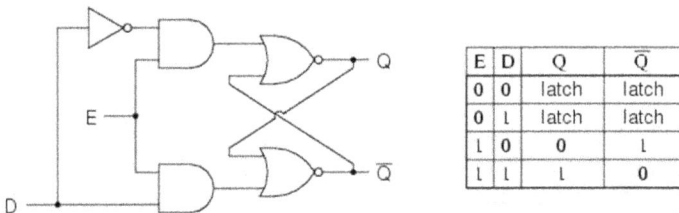

E	D	Q	$\overline{Q}$
0	0	latch	latch
0	1	latch	latch
1	0	0	1
1	1	1	0

Figure 8.13: D Latch.

Note that the R input has been replaced with the complement (inversion) of the old S input, and the S input has been renamed to D. As with the gated S-R latch, the D latch will not respond to a signal input if the enable input is 0—it simply stays latched in its last state. When the enable input is 1, however, the Q output follows the D input.

Since the R input of the S-R circuitry has been done away with, this latch has no "invalid" or "illegal" state. Q and not-Q are *always* opposite of one another. If the above diagram is confusing at all, the next diagram should make the concept simpler (Figure 8.14):

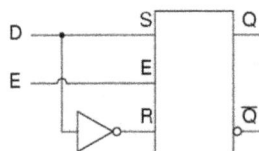

Figure 8.14: S-R Circuitry.

Like both the S-R and gated S-R latches, the D latch circuit may be found as its own prepackaged circuit, complete with a standard symbol (Figure 8.15):

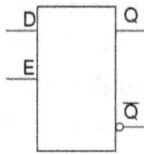

Figure 8.15: D Latch.

The D latch is nothing more than a gated S-R latch with an inverter added to make R the complement (inverse) of S (Figure 8.16). Let's explore the ladder logic equivalent of a D latch, modified from the basic ladder diagram of an S-R latch:

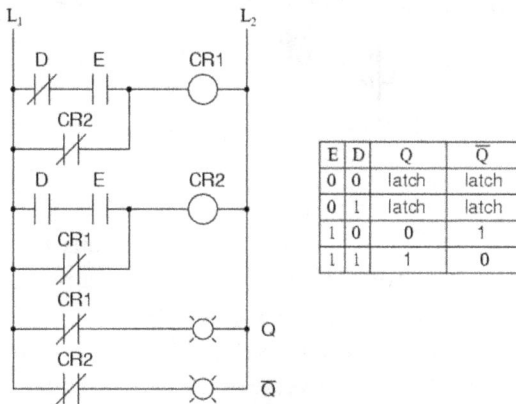

E	D	Q	$\bar{Q}$
0	0	latch	latch
0	1	latch	latch
1	0	0	1
1	1	1	0

Figure 8.16: D Latch.

An application for the D latch is a 1-bit memory circuit. You can "write" (store) a 0 or 1 bit in this latch circuit by making the enable input high (1) and setting D to whatever you want the stored bit to be. When the enable input is made low (0), the latch ignores the status of the D input and merrily holds the stored bit value, outputting at the stored value at Q, and its inverse on output not-Q.

Flip Flops

A flip–flop is a "bit box"; it stores a single binary bit. By Q(t), we denote the state of the flip–flop at the present time, or present tick of the clock; either Q(t) = 0 or Q(t) = 1. The student will note that throughout this textbook we make the assumption that all circuit elements function correctly so that any binary device is assumed to have only two states.

A flip-lop must have an output; this is called either Q or Q(t). This output indicates the current state of the flip–flop, and as such is either a binary 0 or a binary 1. We shall see that, as a result of the way in which they are constructed, all flip–flops also output $\overline{Q(t)}$, the complement of the current state. Each flip–flop also has, as input, signals that specify how the next state, $Q(t+1)$, is to relate to the present state, $Q(t)$.

Every flip-flop also has an input derived from the system clock, which allows it to function as a synchronous circuit. It also has connections to power and ground.

Here are the symbols for SR and D flip-flops, which we have not yet defined. Note the small triangles on the clock input; this says "edge- triggered".

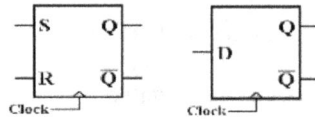

One should note that the more common symbols for flip-flops show the clock input coming in "from the side" as shown in the next figure.

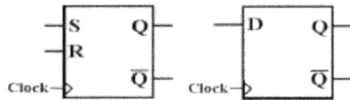

The Clock

The most fundamental characteristic of synchronous sequential circuits is a system clock. This is an electronic circuit that produces a repetitive train of logic 1 and logic 0 at a regular rate, called the **clock frequency**. Most computer systems have a number of clocks, usually operating at related frequencies; for example –2 GHz, 1GHz, 500MHz and 125MHz. The inverse of the clock frequency is the **clock cycle time**. As an example, we consider a clock with a frequency of 2 GHz ($2 \cdot 10^9$ Hertz). The cycle time is $1.0 / (2 \cdot 10^9)$ seconds, or $0.5 \cdot 10^{-9}$ seconds $= 0.500$ nanoseconds $= 500$ picoseconds.

Synchronous sequential circuits are sequential circuits that use a **clock input** to order events. Asynchronous sequential circuits do not use a common clock and, as hinted at above, are much harder to design and test. As we shall focus only on synchronous circuits, we immediately launch a discussion of the clock. The Figure 8.17 illustrates some of the terms commonly used for a clock.

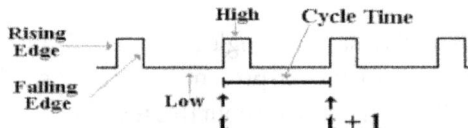

Figure 8.17: Clock Cycle.

The **clock input** is very important to the concept of a sequential circuit. At each "tick" of the clock, the output of a sequential circuit is determined by its input and by its state. We now provide a common definition of a "**clock tick**" – it occurs at the

rising edge of each pulse. We use t to represent the time at a clock tick and (t + 1) to denote the time at the next clock tick – the difference between the two is the **clock cycle time**. Suppose a 2 GHz clock, which corresponds to a clock cycle time of 0.5 nanoseconds. Strictly speaking, we should label our timings in nanoseconds: 1.0, 1.5. 2.0. 2.5, *etc.* The convention is just to count the ticks, referring to the present clock pulse as occurring at time **t** and the next one at a time (**t** + 1).

Description

By definition, a flip-flop is an edge-triggered latch. We must now show to achieve edge triggering. In essence, what we shall do is take a level triggered latch and give it a clock pulse with a very short positive (logical 1) phase.

The key component of an edge-triggered flip-flop is a pulse generator that could be based on the gate delay of a NOT gate. More modern devices likely use a different circuit so as to generate a shorter pulse (Figure 8.18).

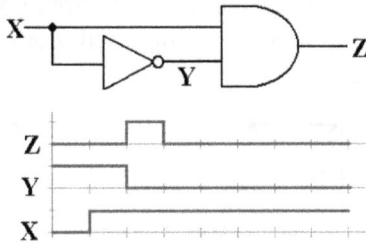

Figure 8.18: Edge–Triggered Latch.

Now that we have a method to generate a short pulse, we can build an edge-triggered device. The Figure 8.19 showing components of two typical edge-triggered flip-flops. The top one is a SR flip-flop and the bottom one is a D flip–flop.

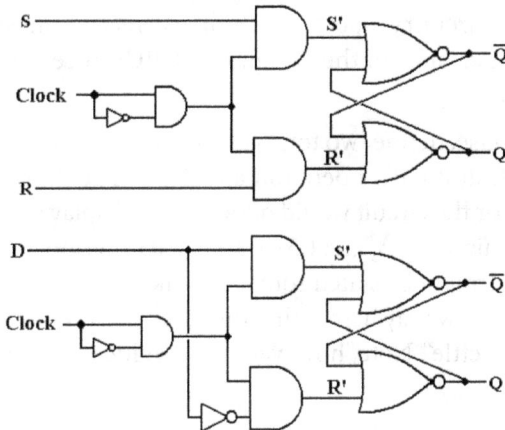

Figure 8.19: Components of Two Typical Edge-Triggered Flip-Flops.

Types of Flip–Flops

In order to avoid the simple example, we shall examine the general case. Here is the circuit for consideration. We postulate a total of gate delays (including that of the D flip-flop) to be the time interval signified by Δ. Thus at time D after the D latch is first sensitive to its input, the circuitry has output a new value to become D. But the flip–flop is activated by a short pulse, of time length δ.

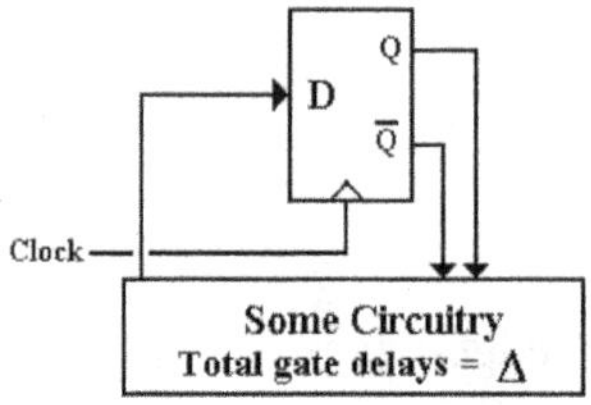

The timing diagram (Figure 8.20) shows the interruption of the uncontrolled feedback loop. By the time that the output of the circuitry has changed, the D flip-flop is no longer sensitive to input. The input will not become effective until the beginning of the next clock cycle.

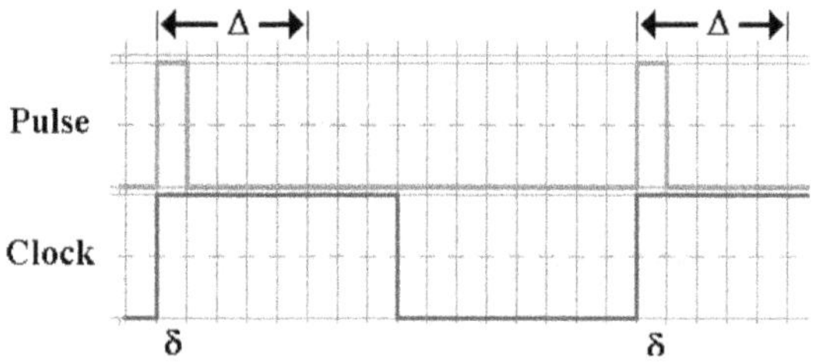

Figure 8.20: Timing Diagram.

How can one ensure that the relative timings of two circuits, the D flip-flop and the rest of the CPU circuitry, operate with the correct timings? This is one of the issues of central importance in the design of a CPU; since we have stumbled into it, let's talk about it.

To be more precise, define two total gate delays: Δ_{MIN} and Δ_{MAX}. Δ_{MIN} is the total time delay for the fastest CPU operation and Δ_{MAX}. The delay for the slowest. We must have $\Delta_{MIN} > \delta$, or the circuit would occasionally display uncontrolled feedback. Conservatively, we might say $\Delta_{MIN} \geq 1.5 \cdot \delta$. The next criterion is a bit more difficult to state precisely, but it might be stated something like $T \geq 1.5 \cdot \Delta_{MAX}$, where T denotes the clock period. What we say here is that the clock period must be long enough for the CPU output to "settle". Note, however, that a value of T much larger than Δ_{MAX} is just wasted time.

Put another way, the value of Δ_{MAX} determines the fastest clock that can reasonably be applied to the CPU. A good part of the art of CPU design is based on this issue. When we study the RISC (Reduced Instruction Set Computer) movement, we shall see that one of the issues was to remove the more complex instructions, thus reducing Δ_{MAX} and allowing for a faster clock. There is a trade-off here that we shall explore in later chapters.

To be honest, the sum of CPU gate delays is not always the limiting factor in the clock speed. Some recent CPUs have been designed with a "hot clock" that is hot in both ways. It is very fast, being of the order of 4 to 5 GHz. It is also hot in the literal sense, in that the CPU, operating at that clock rate, emits so much heat that it overheats itself. The art of CPU design is always bumping up against those messy laws of physics.

SR Flip–Flop

We have completely defined the SR flip-flop, based on the idea of an SR latch. While we could just leave the topic and proceed, it is appropriate at this time to review the subject and state, in one place, what the student is expected to know. Depiction of the SR flip-flop is given in Figure 8.21.

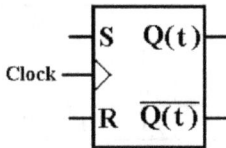

Figure 8.21: SR Flip–Flop.

At this point we are no longer interested in the internal construction of the flip-flop, but on its operational characteristics. These are given in the **characteristic table**.

S	R	$Q(t+1)$
0	0	$Q(t)$
0	1	0
1	0	1
1	1	ERROR

We next address an issue that commonly arises in the use of flip-flops in circuit design. We have a number of scenarios. For each, we know $Q(t)$ and what $Q(t+1)$ should be. The question is how to achieve that change. For example, if $Q(t) = 0$ and we want $Q(t+1) = 0$, we have two choices: either $S = 0$ and $R = 0$, or $S = 0$ and $R = 1$. The first option keeps the state unchanged at 0; the second forces it to 0. As $S = 0$ is sufficient to do this without regard to the value of R, we say that the input is $S = 0$ and $R = d$; the d standing for "don't care".

On the other hand, if $Q(t) = 0$ and $Q(t+1) = 1$, only $S = 1$ and $R = 0$ will do. This is the only combination that will give the next state of 1 when the present state is 0.

If we have $Q(t) = 1$ and want $Q(t+1) = 0$, then our choice is simple: $S = 0$ and $R = 1$.

If we have $Q(t) = 1$ and want $Q(t + 1) = 1$, then we can choose either $S = 0$ and $R = 0$, or

$S = 1$ and $R = 0$. This is denoted as $S = d$ and $R = 0$.

The above discussions lead to the **excitation table** for the SR flip–flop.

Q(t)	Q(t + 1)	S	R
0	0	0	d
0	1	1	0
1	0	0	1
1	1	D	0

JK Flip–Flop: Enhancing SR Flip–Flop

Recall the characteristic table of the SR flip-flop. We repeat the table here for emphasis.

S	R	Q(t + 1)
0	0	Q(t)
0	1	0
1	0	1
1	1	ERROR

The theoretician examining this table would note two facts immediately.

1. The input $S = 1$ and $R = 1$ is disallowed; we would like to do something with it.

2. The values for $Q(t + 1)$ are three of the possible four Boolean functions of 1 variable.

Considering Q as a Boolean variable, we now show that there are exactly four Boolean functions of this Boolean variable. These are $f(Q) = 0$, $f(Q) = 1$, $f(Q) = Q$, and $f(Q) = \overline{Q}$. We do this by showing the truth table for each of these functions and noting that there are only four different ways to put 0's and 1's into the two row entries for a function.

Q	0	Q	$\overline{Q}$	1
0	0	0	1	1
1	0	1	0	1

The Four Boolean Functions of Boolean Variable Q

Given this, our enhanced SR flip-flop would have $Q(t + 1) = \overline{Q}(t)$ as one possible output. For maximal compatibility with the existing SR, we would want to leave the existing valid inputs alone and just make good use of the invalid one. What we get is a JK flip–flop.

Recalling that an SR flip-flop is so called because it is a **Set-R**eset device, we may ask for the meaning of JK. In any case, we present the characteristic table for the JK flip–flop and then ask how one might modify an SR flip to achieve that goal.

Here is the desired characteristic table.

J	K	$Q(t+1)$
0	0	$Q(t)$
0	1	0
1	0	1
1	1	$\overline{Q}(t)$

Viewing this as a modification of the SR flip–flop, we ask how to generate each of S and R from the inputs J and K under the following constraints;

1. Except when $J = 1$ and $K = 1$, we want to have $S = J$ and $R = K$.

   Under these circumstances, the behavior is identical.

2. When $J = 1$, $K = 1$ and $Q = 0$, we want $S = 1$ and $R = 0$. This makes $Q(t+1) = 1$.

3. When $J = 1$, $K = 1$ and $Q = 1$, we want $S = 0$ and $R = 1$. This makes $Q(t+1) = 0$.

When in doubt about how to create a circuit, we make a truth table. The inputs to the truth table are J, K and Q (the present state). The outputs are S and R.

Row	Q	J	K	S	R
0	0	0	0	0	0
1	0	0	1	0	1
2	0	1	0	1	0
3	0	1	1	1	0
4	1	0	0	0	0
5	1	0	1	0	1
6	1	1	0	1	0
7	1	1	1	0	1

We have two patterns that almost work: $S = J \cdot \overline{Q}$ and $R = K \cdot Q$. Let's examine each case. $S = J \cdot \overline{Q}$ produces the expected result for all rows except row 6. In that row we have $Q = 1$, $J = 1$ and $K = 0$. We want $Q(t+1)$ to be 1. But $S = 0$ and $R = 0$ will cause $Q(t+1) = Q(t) = 1$, exactly what we want. So this simple formula causes no trouble.

$R = K \cdot Q$ produces the expected result for all rows except row 1. In that row we have $Q = 0$, $J = 0$ and $K = 1$. We want $Q(t+1)$ to be 0. But $S = 0$ and $R = 0$ will cause $Q(t+1) = Q(t) = 0$, exactly what we want.

Representation of JK flip-flop is given in Figure 8.22.

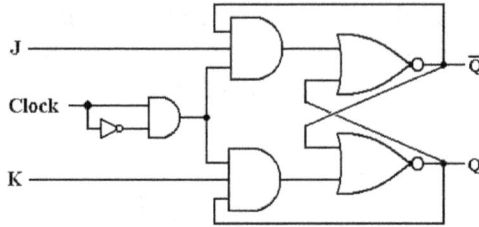

Figure 8.22: JK Flip–Flop.

We now give the standard representation of the JK flip–flop as will be used in future discussions. Again, note the triangle symbol on the clock input, indicating edge triggering (Figure 8.23).

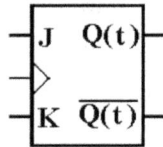

Figure 8.23: Clock of JK Flip–Flop.

We have already presented the **characteristic table** for the JK flip–flop. Here it is again.

J	K	$Q(t+1)$
0	0	$Q(t)$
0	1	0
1	0	1
1	1	$\overline{Q}(t)$

We now create the excitation table for the JK.

If $Q(t) = 0$ and $Q(t + 1)$ is to be 0, we can use either $J = 0$ and $K = 0$, or $J = 0$ and $K = 1$.

If $Q(t) = 0$ and $Q(t + 1)$ is to be 1, we can use either $J = 1$ and $K = 0$, or $J = 1$ and $K = 1$. Note that this is a new option, not available for an SR flip–flop.

If $Q(t) = 1$ and $Q(t + 1)$ is to be 0, we can use either $J = 0$ and $K = 1$, or $J = 1$ and $K = 1$. Note that this also is a new option, not available for an SR flip–flop.

If $Q(t) = 1$ and $Q(t + 1)$ is to be 1, we can use either $J = 0$ and $K = 0$, or $J = 1$ and $K = 0$.

This gives the **excitation table** for the JK flip–flop.

Q(t)	Q(t + 1)	J	K
0	0	0	d
0	1	1	d
1	0	d	1
1	1	d	0

D Flip–Flop

We have already examined the D latch. The input is labeled "D" for "Data". The D flip–flop has a characteristic table identical to that of a D latch.

D	Q(t + 1)
0	0
1	1

The device is so simple that it does not require an excitation equation. We just use an excitation equation, which simply states "Give it what you want".

$$D = Q(t + 1)$$

T Flip–Flop

This is the fourth and last of the major types of flipflops. The input to this flipflop is labeled "T" for toggle. When T = 0, the state remains the same. When T = 1, the flip–flop changes state. This gives rise to the following **characteristic table**.

T	Q(t + 1)
0	Q(t)
1	$\overline{Q}(t)$

The **excitation table** for this flip–flop is almost obvious.

Q(t)	Q(t + 1)	T
0	0	0
0	1	1
1	0	1
1	1	0

This gives rise to an excitation equation.

$$T = Q(t) \oplus Q(t + 1)$$

JK as General Flip–Flop

We now take notice that the JK can be configured to function as any of the other 3 flip–flop types. In this, it is the most general type of flip–flop.

JK as D Flip–Flop

To convert a JK to a D flip–flop, connect the inputs as follows (Figure 8.24)

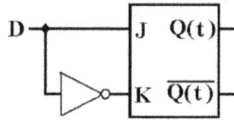

Figure 8.24: JK to a D Flip–Flop.

If $D = 0$, then $J = 0$, $K = 1$ and $Q(t+1) = 0$, If $D = 1$, then $J = 1$, $K = 0$ and $Q(t+1) = 1$.

JK as T FlipFlop

To convert a JK to a T flip–flop, connect the inputs as follows (Figure 8.25)

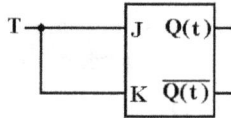

Figure 8.25: JK to T Flip–Flop.

If $T = 0$, then $J = 0$, $K = 0$ and $Q(t+1) = Q(t)$.

If $T = 1$, then $J = 1$, $K = 1$ and $Q(t+1) = \overline{Q}(t)$

Analysis of Clocked Sequential Circuits

The analysis consists of obtaining a state-table or a state-diagram from a given sequential circuit implementation. In other words, analysis closes the loop by forming state-table from a given circuit-implementation. We will show the analysis procedure by deriving the state table of the example circuit we considered in synthesis.

The circuit is shown in Figure 8.26.

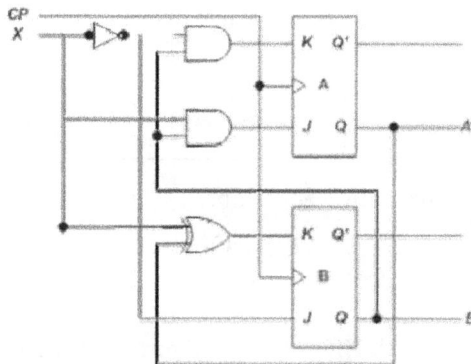

Figure 8.26: A Clocked Sequential Circuit.

A State table is a representation of a sequence of inputs, outputs, and flip-flop states in a tabular form. Two forms of state tables are shown.

Present state		Next State				Output	
		x = 0		x = 1		x = 0	x = 1
A	B	A	B	A	B	y	y
0	0						
0	1						
1	0						
1	1						

State Table: Form 1

The analysis is the generation of state table from the given sequential circuit.

The number of rows in the state table is equal to 2 $^{\text{(number of flip-flops+ number of inputs)}}$. For the circuit under consideration, number of rows $= 2^{(2+1)} = 2^{(3)} = 8$

Present state		Input	Next state		Output
A(t)	B(t)	X	A(t+1)	B(t+1)	y
0	0	0			
0	0	1			
0	1	0			
0	1	1			
1	0	0			
1	0	1			
1	1	0			
1	1	1			

State Table: Form 2

In the present case there are two flip-flops and one input, thus a total of 8 rows as shown in the table.

Present state		Input	Next state		Output
A(t)	B(t)	X	A(t+1)	B(t+1)	y
0	0	0	0	0	0
0	0	1	0	1	0
0	1	0	1	0	0
0	1	1	0	1	0
1	0	0	0	0	0
1	0	1	1	1	0
1	1	0	1	0	0
1	1	1	0	1	1

State Table

The analysis can start from any arbitrary state. Let us start deriving the state table from the initial state 00. As a first step, the input equations to the flip-flops and to the combinational circuit must be obtained from the given logic diagram. These equations are:

$$J_A = BX'$$

$K_A = BX + B'X'$

$D_B = X$

$y = ABX$

The first row of the state-table is obtained as follows:

When input $X = 0$; and present states $A = 0$ and $B = 0$ (as in the first row); then, using the above equations we get:

$y = 0$, $J_A = 0$, $K_A = 1$ and $D_B = 0$.

The resulting state table is exactly same from which we started our design example. This analysis is opposite to design and combined they act as a closed loop.

State Reduction

The problem of state reduction is to find ways of reducing the number of states in a sequential circuit without altering the input-output relationships.

In other words, to reduce the number of states, redundant states should be eliminated. A redundant state S_i is a state which is equivalent to another state S_j.

Two states are said to be equivalent if, for each member of the set of inputs, they give exactly the same output and send the circuit either to the same state or to an equivalent state.

Since 'm' flip-flops can describe a state machine of up to 2^m states, reducing the number of states may (or may not) result in a reduction in the number of flip-flops. For example, if the number of states is reduced from 8 to 5, we still need 3 flip-flops. However, state reduction will result in more don't care states. The increased number of don't care states can help obtain a simplified circuit for the state machine.

Consider the shown state diagram in Figure 8.27.

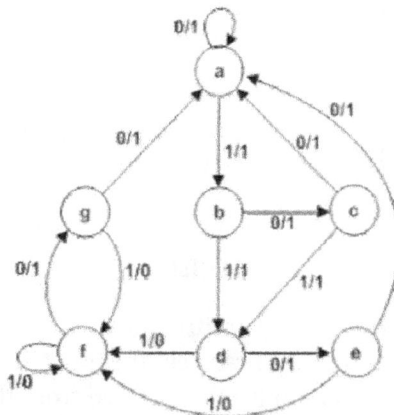

Figure 8.27: State Diagram.

The state reduction proceeds by first tabulating the information of the state diagram into its equivalent state-table form (as shown in the table). The problem of state reduction requires identifying equivalent states. Each N states is replaced by 1 state.

Consider the following state table.

Present state	Next state		Output	
	$x=0$	$x=1$	$x=0$	$x=1$
a	a	b	1	1
b	c	d	1	1
c	a	d	1	1
d	e	̶f̶d	1	0
e	a	̶f̶d	1	0
̶f̶	̶g̶e	f	1	0
̶g̶	a	f	1	0

State Table after Reduction

States 'g' and 'e' produce the same outputs, *i.e.* '1' and '0', and take the state machine to same next-states, 'a' and 'f', on inputs '0' and '1' respectively. Thus, states 'g' and 'e' are equivalent states.

We can now remove state 'g' and replace it with 'e' as shown. We next note that the above change has caused the states 'd' and 'f' to be equivalent. Thus in the next step, we remove state 'f' and replace it with 'd'. There are no more equivalent states remaining. The reduced state table results in the following reduced state diagram.

States Assignment

When constructing a state diagram, variable names are used for states as the final number of states is not known a priori. Once the state diagram is constructed, prior to implementation (using gates and flip-flops), we need to perform the step of 'state reduction'. The step that follows state reduction is state assignment. In state assignment, binary patterns are assigned to state variables.

State	Assignment 1	Assignment 2	Assignment 3
a	001	000	000
b	010	010	100
c	011	011	010
d	100	101	101
e	101	111	011

Possible State Assignments

For a given machine, there are several state assignments possible. Different state assignments may result in different combinational circuits of varying complexities. State assignment procedures try to assign binary values to states such that the cost (*complexity*) of the combinational circuit is reduced. There are several heuristics that attempt to choose good state assignments (also known as state encoding) that try to reduce the required combinational logic complexity, and hence cost.

As mentioned earlier, for the reduced state machine obtained in the previous example, there can be a number of possible assignments. As an example, three different state assignments are shown in the table for the same machine.

Design Procedure

There are occasions when a sequential circuit, implemented using m flip-flops, may not utilize all the possible 2^m states

Present state	Next state		Output	
	$x = 0$	$x = 1$	$x = 0$	$x = 1$
a	a	b	1	1
b	c	d	1	1
c	a	d	1	1
d	e	d	1	0
e	a	d	1	0

Reduced Table with Binary Assignments

In the previous example of a machine with 5 states, we need three flip-flops. Let us choose assignment 1, which is binary assignment for our sequential machine example (shown in the table). The unspecified states can be used as don't-cares and will therefore, help in simplifying the logic. The excitation table of the previous example is shown. There are three states, 000, 110 and 111 that are not listed in the table under present state and input.

Present state			Input	Next state			Flip-flop inputs						Output
A	B	C	x	A	B	C	SA	RA	SB	RB	SC	RC	y
0	0	1	0	0	0	1	0	X	0	X	X	0	1
0	0	1	1	0	0	1	0	X	1	0	0	1	1
0	1	0	0	0	1	0	0	X	X	0	1	0	1
0	1	0	1	0	1	0	1	0	0	1	0	X	1
0	1	1	0	0	1	1	0	X	0	1	X	0	1
0	1	1	1	0	1	1	1	0	0	1	0	1	1
1	0	0	0	1	0	0	X	0	0	X	1	0	1
1	0	0	1	1	0	0	X	0	0	X	0	X	0
1	0	1	0	1	0	1	0	1	0	X	X	0	1
1	0	1	1	1	0	1	X	0	0	X	0	1	0

Excitation Table

With the inclusion of input 1 or 0, we obtain six don't-care minterms: 0, 1, 12, 13, 14 and 15.

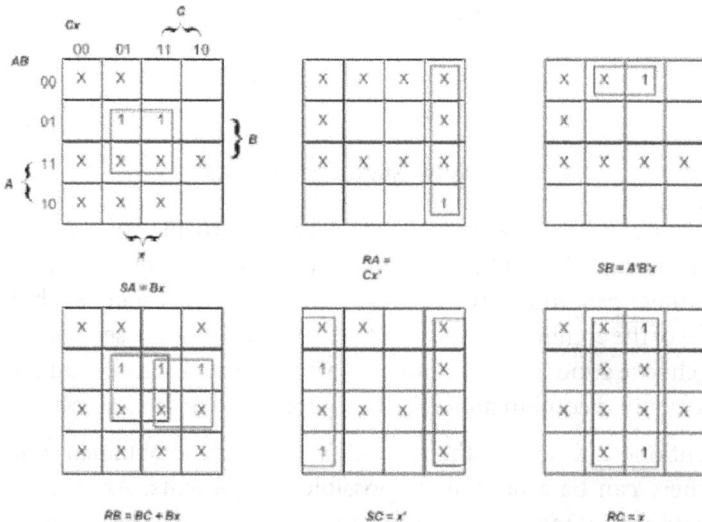

SA = Bx

RA = Cx'

SB = A'B'y

RB = BC + Bx

SC = x'

RC = y

The K-maps of SA and RA is shown in the figure. Other K-Maps can be obtained similarly and the equations derived are shown in the Figure 8.28.. The logic diagram thus obtained is shown in the Figure 8.28.

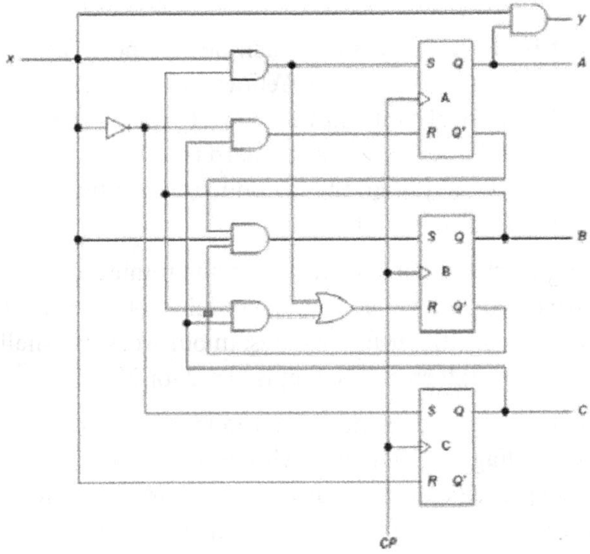

Figure 8.28: Logic Diagram.

$SA = Bx$

$RA = Cx'$

$SB = A'B'x$

$RB = BC + Bx$

$SC = x'$

$RC = x$

$y = A' + x' = (Ax)'$

Note that the design of the sequential circuit is dependent on binary codes for states. A different binary state codes set may have resulted in some different combinational circuit.

Finite State Machines (FSM)

Strictly speaking a **finite state machine** (FSM) is a device that can exist in one of a finite number of states. Associated with an FSM is a memory that is used to store an identifier of the state, so that the machine may process its input (if any) and move to the next state. Due to this coincidence, finite state machines are often studied in conjunction with flip-flops.

We are all familiar with finite state machines, although we rarely think of them as such. Consider a washing machine. The states for this machine are: off, fill with water, wash, spin and rinse. The control unit for the FSM is the knob on the washer that we turn to start it.

A traffic light is also a finite state machine. We normally think of a traffic light as having only three states: Green, Yellow and Red. The truth is a bit more complex, in that the physical unit must display at least two sets of lights, one for each intersecting street. Nevertheless, a standard traffic light can be modeled with no more than eight states, although the introduction of advanced green lights and turn signals complicates things a bit.

A standard digital clock that displays only hour, minute and second, can be said to have $24 \cdot 60 \cdot 60 = 86,400$ states – still a finite number. Normally the FSM construct is used to model systems with far fewer states; in our work, we shall normally limit a FSM to either eight or sixteen states; that is $N \leq 2^3$ or $N \leq 2^4$.

If a finite state machine does not have too many states, we may represent its operation by a **state diagram**. The following is a state diagram (Figure 8.29) for a circuit called a sequence detector. For those who are interested, this is the state diagram for a 11011 sequence detector; it has five states because it is detecting a five-bit sequence.

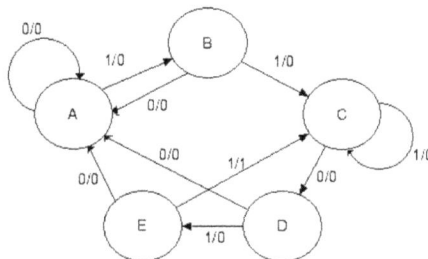

Figure 8.29: State Diagram for a 11011 Sequence Detector.

At this stage of the presentation, we focus not on the details of generating the state diagram, but just use it as an example of a generic state diagram. What do we note about this one?

1. In terms of discrete mathematics, it is a directed graph with loops. Thus, it is not a simple graph. In simple graphs, arcs do not connect any vertex with itself.

2. The arcs each have direction and a label of the form X/Z. What we see here is the FSM reacting to input by moving between states and producing output. In the X/Z labeling scheme, X is the binary input (0 or 1) and Z is the binary output.

3. There is output associated with the transitions. Not all FSM have output associated with the transitions between states. This one does.

4. This and all typical FSM represents a synchronous machine; transitions between states and production of output (if any) takes place at a fixed phase of the clock, depending on the flip-flops used to implement the circuit.

Not all finite state machines have such complex state diagrams.

The figure is the state diagram for a modulo-4 up-counter (Figure 8.30). It just counts 0, 1, 2, 3 and repeats, continually counting up (modulo 4). There is no input (other than the clock, which we almost never mention) and no output directly associated with the transitions. For this type of FSM, the output is associated with the states and not with the transitions.

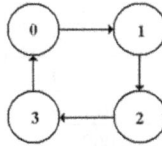

Figure 8.30: State Diagram for a Modulo-4 Counter..

Many mathematical models of FSM focus on the state diagram. For most of our work, it is more convenient to work with the **state table** of the FSM, a tabular representation of the state diagram. Translation between the state diagram and state table is automatic. The state table presents the data in terms of **present state** $Q(t)$ and **next state** $Q(t+1)$ using the labeling that most naturally fits the problem. Here are the state tables for the two FSM above. Note that eh state table contains exactly the same information as the state diagram.

Present State	Next State / Output	
	$X = 0$	$X = 1$
A	A / 0	B / 0
B	A / 0	C / 0
C	D / 0	C / 0
D	A / 0	E / 0
E	A / 0	C / 1

State Table for 11011 Sequence Detector, Showing Output

Present State	Next State
0	1
1	2
2	3
3	0

State Table for a Modulo-Four Up-Counter

One notes immediately that the second state table is simpler than the first; this is expected it represents a simpler state diagram. Specifically, there is no input, so there is only one column for the next state. In general, for K inputs there are 2^K next state columns in the table.

Another tool in the design and analysis of sequential circuits is the **transition table**. It contains the same information as the state table, except that all labels have been replaced by binary numbers. There are many creative ways to assign binary numbers to state labels, here we just do the obvious. For the sequence detector, let $A = 000, B = 001, C = 010, D = 011$ and $E = 100$ (as there are five states). The following is the sequence detector transition table.

Present State	Next State / Output	
	X = 0	X = 1
A = 000	000 / 0	001 / 0
B = 001	000 / 0	010 / 0
C = 010	011 / 0	010 / 0
D = 011	000 / 0	100 / 0
E = 100	000 / 0	010 / 1

Transition/Output Table for the 11011 Sequence Detector

What we have in the above figure is a special type of truth table. We shall now investigate the table in a bit more detail. Note that the state of the machine is represented by a 3-bit binary number. We shall use the notation $Y_2Y_1Y_0$ for that number. Given this notation, we write the table as shown below.

Present			Next State/Output							
State			X = 0				X = 1			
Y_2	Y_1	Y_0	Y_2	Y_1	Y_0	Z	Y_2	Y_1	Y_0	Z
0	0	0	0	0	0	0	0	0	1	0
0	0	1	0	0	0	0	0	1	0	0
0	1	0	0	1	1	0	0	1	0	0
0	1	1	0	0	0	0	1	0	0	0
1	0	0	0	0	0	0	0	1	0	1

Transition/Output Table as Modified Truth-Table

The table above can be viewed as a truth table that has been "folded over". Another way to represent this table is as a standard truth table depending on Y_2, Y_1, Y_0, and X.

Y_2	Y_1	Y_0	X	Y_2	Y_1	Y_0	Z
0	0	0	0	0	0	0	0
0	0	0	1	0	0	1	0
0	0	1	0	0	0	0	0

Y_2	Y_1	Y_0	X	Y_2	Y_1	Y_0	Z
0	0	1	1	0	1	0	0
0	1	0	0	0	1	1	0
0	1	0	1	0	1	0	0
0	1	1	0	0	0	0	0
0	1	1	1	1	0	0	0
1	0	0	0	0	0	0	0
1	0	0	1	0	1	0	1

Transition/Output Table as a Standard Truth Table

We now have three equivalent representations of a FSM.

1. the state diagram,
2. the state table, and
3. the transition/output table (probably not a standard name).

Design of Sequential Circuits

Having seen how to analyze digital circuits, we now investigate how to design digital circuits. We assume that we are given a complete and unambiguous description of the circuit to be designed as a starting point. At this level, most design problems focus on one of two topics: modulo-N counters and sequence detectors. Here is an overview of the design procedure for a sequential circuit.

1. Derive the state diagram and state table for the circuit.
2. Count the number of states in the state diagram (call it N) and calculate the number of flip-flops needed (call it P) by solving the equation $2^{P-1} < N \leq 2^P$. This is best solved by guessing the value of P.
3. Assign a unique P-bit binary number (state vector) to each state. Often, the first state = 0, the next state = 1, *etc.*
4. Derive the state transition table and the output table.
5. Separate the state transition table into P tables, one for each flip-flop. WARNING: Things can get messy here; neatness counts.
6. Decide on the types of flip-flops to use. When in doubt, use all JK's.
7. Derive the input table for each flip-flop using the excitation tables for the type.
8. Derive the input equations for each flip-flop based as functions of the input and the current state of all flip-flops.
9. Summarize the equations by writing them in one place.
10. Draw the circuit diagram. Most homework assignments will not go this far, as the circuit diagrams are hard to draw neatly.

Modulo-4 Counter

As our first design problem, let's consider a modulo-four counter. When the direction is not specified, we usually intend to build a modulo-four up-counter: 0, 1, 2, 3, 0, 1, 2, 3, *etc.* We solve these design problems by using the step-wise procedure listed above.

Step 1: Derive the state diagram and state table for the circuit.

Here is the state diagram. Note that it is quite simple and involves no input.

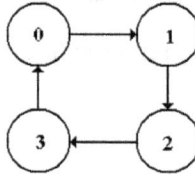

Figure 8.31: State Diagram for a Modulo-4 Up Counter.

The state table is simply a rearrangement of the state diagram into a tabular form.

Present State	Next State
0	1
1	2
2	3
3	0

Step 2: Count the number of states in the state diagram (call it N) and calculate the number of flip-flops needed (call it P) by solving the equation $2^{P-1} < N \leq 2^P$.

The number of states on a modulo-N counter is simply N; these are labeled 0 through (N – 1). Specifically, a modulo-4 counter has four states: labeled 0, 1, 2, and 3.

We solve the equation $2^{P-1} < 4 \leq 2^P$ by noting that $2^1 = 2$ and $2^2 = 4$, so we have determined that $2^1 < 4 \leq 2^2$, hence P = 2. We shall see later that there are valid solutions with more than two flip-flops, but there are none with fewer.

Step 3: Assign a unique P-bit binary number (state vector) to each state. Often, the first state = 0, the next state = 1, *etc.*

Some sequential circuits suggest an innovative numbering system, but modulo-N counters never do. We go with the obvious labeling, generated by assigning each decimal number its two-bit binary equivalent as an unsigned integer in the range from 0 to 3.

State	2-bit Vector
0	0 0
1	0 1
2	1 0
3	1 1

Step 4: Derive the state transition table and the output table.

There is no output table for any modulo-N counter, as the output associated with this type of table is the output on a transition, as is seen in a sequence detector. The transition table is a direct translation of the state table, using the assignments from the previous step.

Present State		Next State
0	00	01
1	01	10
2	10	11
3	11	00

Step 5: Separate the state transition table into P tables, one for each flip-flop.

Here we separate the state transition table into 2 tables, one for each flip-flop. Note that the flip-flops will be numbered 0 and 1, with flip-flop 0 storing the least significant bit of the state information. Thus, we shall refer to the state information as $Y_1 Y_0$.

Flip-Flop 1			Flip-Flop 0	
Present State	Next State		Present State	Next State
$Y_1 Y_0$	$Y_1(T+1)$		$Y_1 Y_0$	$Y_0(T+1)$
0 0	0		0 0	1
0 1	1		0 1	0
1 0	1		1 0	1
1 1	0		1 1	0

Step 6: Decide on the types of flip-flops to use. When in doubt, use all JK's.

Up to this point, we have made no assumptions about the type of flip-flop to use. In order to proceed any farther with the design, we must now commit to a specific type. In line with this author's preferences, he chooses to use two JK flip-flops.

The excitation table for a JK flip-flop is shown at right. Recall that the "d" stands for "Don't Care". For example, if we have $Q(t) = 0$ and want $Q(t+1) = 0$, we can use either $J = 0, K = 0$ or $J = 0, K = 1$. Similarly either $J = 1$ and $K = 0$ or $J = 1$ and $K = 1$ will take $Q(t) = 0$ to $Q(t + 1) = 1$.

$Q(t)$	$Q(t+1)$	J	K
0	0	0	d
0	1	1	d
1	0	d	1
1	1	d	0

Step 7: Derive the input table for each flip-flop using the excitation tables for the type.

First look at flip-flop 1, representing the high-order bit. Note that we compare the present state of Y1 to its next state in order to determine J1 and K1.

PS	NS	Input	
Y_1 Y_0	Y_1	J_1	K_1
0 0	0	0	d
0 1	1	1	d
1 0	1	d	0
1 1	0	d	1

Note that in deciding on the input, we must match only the 0's and 1's. We ignore the don't-cares. Note that the "d" for "don't-care" is not a variable to be assigned a value. It is a value that does not need to be matched. At the moment, Y_0 is included in the table for future use only. It plays no part in determining the values of J_1 and K_1.

Here is the table for Y_0

PS	NS	Input	
Y_1 Y_0	Y_0	J_0	K_0
0 0	1	1	d
0 1	0	d	1
1 0	1	1	d
1 1	0	d	1

Again, Y_1 is included in the table for future use only. It plays no part in determining the values of J_0 and K_0.

Step 8: Derive the input equations for each flip-flop based as functions of the input and current state of all flip-flops.

At this point, we try to derive an expression that matches each column. Formal methods can be used, but generally are more trouble than they are worth. Here is this author's set of rules to match an expression to a given column.

1. If a column does not have a 0 in it, match it to the constant value 1. If a column does not have a 1 in it, match it to the constant value 0.

2. If the column has both 0's and 1's in it, try to match it to a single variable, which must be part of the present state. Only the 0's and 1's in a column must match the suggested function.

3. If every 0 and 1 in the column is a mismatch, match to the complement of a function.

4. If all the above fails, try for simple combinations of the present state.

Let's look at the input table for Y_1.

PS	NS	Input	
Y_1 Y_0	Y_1	J_1	K_1
0 0	0	0	d
0 1	1	1	d
1 0	1	d	0
1 1	0	d	1

Note that the column for J_1 has a 0 and a 1 in it as does the column for K_1. Each column has two "don't cares" in it, but we ignore these. Because each column has both a 0 and a 1 in it, neither is a match for a constant function. We now try to match J_1.

☆ J_1 does not match Y_1, because Y_1 is 0 in the same row (0 1) as J_1 is 1.

☆ J_1 matches Y_0. In row 0 0, both Y_0 and J_1 are 1. In row 0 1, both Y_0 and J_1 are 1.

☆ In rows 1 0 and 1 1, J_1 is a "don't care", so we do not need to match it.

Similar logic shows that K_1 matches Y_0 also.

So now we have the following matches for J_1 and K_1.

PS	NS	Input	
Y_1 Y_0	Y_1	J_1	K_1
0 0	0	0	d
0 1	1	1	d
1 0	1	d	0
1 1	0	d	1
		$J_1 = Y_0$	$K_1 = Y_0$

We now examine Y_0

PS	NS	Input	
Y_1 Y_0	Y_0	J_0	K_0
0 0	1	1	d
0 1	0	d	1
1 0	1	1	d
1 1	0	d	1

Note that there are no 0's in either the J_0 or K_0 column. The simplest (and best) match is $J_0 = 1$ and $K_0 = 1$.

Step 9: Summarize the equations by writing them in one place.

Here they are.

$$J_1 = Y_0 \qquad K_1 = Y_0$$
$$J_0 = 1 \qquad K_0 = 1$$

This is a counter, so there is no Z output.

Step 10: Draw the circuit diagram.

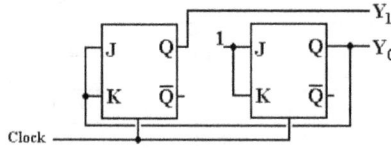

But wait – there is another solution hidden here. Recall that a JK flip-flop can be used to emulate a T flip-flop by setting the J input equal to the K input. Note that the design has the following interesting property.

$$J_1 = K_1 = Y_0$$
$$J_0 = K_0 = 1$$

Given this, we can replace each JK flip-flop with a T flip-flop, arriving at this design.

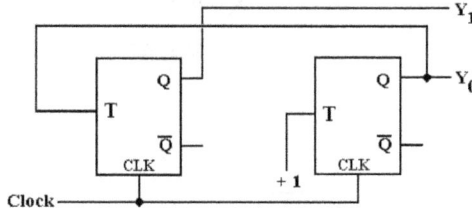

The modulo-4 counter just designed outputs binary codes for the time pulses. Specifically, we assume that it is initialized to $Y_1 Y_0 = 00$ and then outputs 01, 10, 11, 00, 01, 10, 11, *etc.* A more realistic circuit would output discrete pulses corresponding to the decoded output, so that first $T_0 = 1$ and all others are 0, then $T_1 = 1$ and all others are 0, *etc.* In order to produce the discrete signals T_0, T_1, T_2 and T_3, we need to add a decoding phase to the counter.

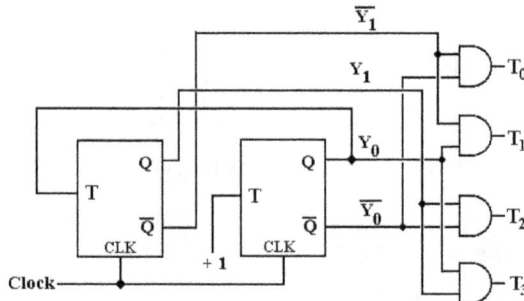

Note that the above design is simplified by the fact that the outputs Y_1' and Y_0' are available directly from the flip-flops and do not need to be synthesized using NOT gates.

One can achieve a simpler design at the cost of additional flip-flops. The following design is called a **one-hot design**, in that it uses a shift register in which exactly one flip-flop at a time is storing a 1. This design also works as a modulo-4 counter and skips the decoder delays (Figure 8.32).

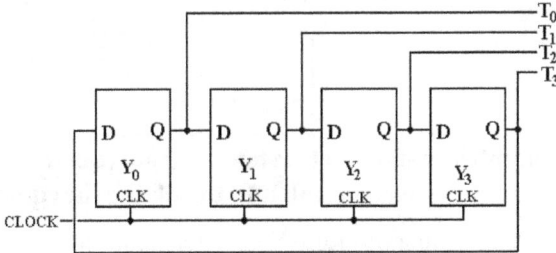

Figure 8.32: Modulo-4 Counter.

When the counter is initialized, we set $Y_0 = 1$ and $Y_1 = Y_2 = Y_3 = 0$. As the clock ticks, the single 1 is shifted by the shift register, so that the discrete signals become high in sequence.

Modulo-4 Up-Down Counter

For the next design, we introduce a problem that uses input. This is a modulo-4 up-down counter. The input X is used to control the direction of counting.

If $X = 0$, the device counts up: 0, 1, 2, 3, 0, 1, 2, 3, *etc.*

If $X = 1$, the device counts down: 0, 3, 2, 1, 0, 3, 2, 1, *etc.*

Step 1: Derive the state diagram and state table for the circuit.

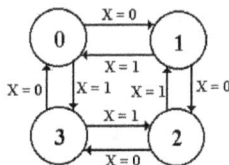

The state diagram for the modulo-4 up-down counter is shown at right. Notice that the X input is used to determine the counting direction. Again, this type of circuit does not have any output associated with the transitions; the output just reflects which of the four states the machine finds itself in at the moment.

We now produce the same table by translating the state diagram. As an aside, some students might prefer to begin the design process with the state table and omit the state diagram. That is certainly acceptable practice; whatever works should be used.

Here, the state table depends on X – the input used to specify the counting direction.

Present State	Next State	
	X = 0	X = 1
0	1	3
1	2	0
2	3	1
3	0	2

Step 2: Count the number of states in the state diagram (call it N) and calculate the number of flip-flops needed (call it P) by solving the equation $2^{P-1} < N \leq 2^P$.

The number of states on a modulo-N counter is simply N; these are labeled 0 through (N – 1). Specifically, a modulo-4 counter has four states: labeled 0, 1, 2 and 3.

We solve the equation $2^{P-1} < 4 \leq 2^P$ by noting that $2^1 = 2$ and $2^2 = 4$, so we have determined that $2^1 < 4 \leq 2^2$, hence P = 2. We shall see later that there are valid solutions with more than two flip-flops, but there are none with fewer.

Step 3: Assign a unique P-bit binary number (state vector) to each state. Often, the first state = 0, the next state = 1, *etc.*

Some sequential circuits suggest an innovative numbering system, but modulo-N counters never do. We go with the obvious labeling, generated by assigning each decimal number its two-bit binary equivalent as an unsigned integer in the range from 0 to 3.

State	2-bit Vector
0	0 0
1	0 1
2	1 0
3	1 1

Step 4: Derive the state transition table and the output table.

There is no output table for any modulo-N counter, as the output associated with this type of table is the output on a transition, as is seen in a sequence detector. The transition table is a direct translation of the state table, using the assignments from the previous step. The transition table for the modulo-4 up-down counter is as follows.

Present State		Next State	
		X = 0	X = 1
0	00	01	11
1	01	10	00
2	10	11	01
3	11	00	10

Step 5 Separate the state transition table into P tables, one for each flip-flop.

Here we separate the state transition table into 2 tables, one for each flip-flop. Note that the flip-flops will be numbered 0 and 1, with flip-flop 0 storing the least significant bit of the state information. Thus, we shall refer to the state information as $Y_1 Y_0$.

Flip-Flop 1				Flip-Flop 0		
PS	Next State			PS	Next State	
$Y_1\ Y_0$	$Y_1, X = 0$	$Y_1, X = 1$		$Y_1 Y_0$	$Y_0, X = 0$	$Y_0, X = 1$
0 0	0	1		0 0	1	1
0 1	1	0		0 1	0	0
1 0	1	0		1 0	1	1
1 1	0	1		1 1	0	0

The student who is paying attention at this point will notice an interesting feature concerning flip-flop 0; specifically that its next state does not depend on X. This is due to the fact that in considering a modulo-N counter, one moves from odd numbers to even numbers and from even numbers to odd numbers in both counting up and counting down.

Step 6: Decide on the types of flip-flops to use. When in doubt, use all JK's.

Up to this point, we have made no assumptions about the type of flip-flop to use. In order to proceed any farther with the design, we must now commit to a specific type.

The excitation table for a JK flip-flop is shown at right. Recall that the "d" stands for "Don't Care". For example, if we have $Q(t) = 0$ and want $Q(t+1) = 0$, we can use either $J = 0, K = 0$ or $J = 0, K = 1$. Similarly either $J = 1$ and $K = 0$ or $J = 1$ and $K = 1$ will take $Q(t) = 0$ to $Q(t + 1) = 1$.

Q(t)	Q(t+1)	J	K
0	0	0	d
0	1	1	d
1	0	d	1
1	1	d	0

Step 7: Derive the input table for each flip-flop using the excitation tables for the type.

Here is the input table for flip-flop 1. Note that the arrangement of the table has been altered to reflect the fact that we now have a binary input.

		X = 0			X = 1	
$Y_1\ Y_0$	Y_1	J_1	K_1	Y_1	J_1	K_1
0 0	0	0	d	1	1	d
0 1	1	1	d	0	0	d
1 0	1	d	0	0	d	1
1 1	0	d	1	1	d	0

Here is the input table for flip-flop 0.

$Y_1\ Y_0$	$X=0$			$X=1$		
	Y_0	J_0	K_0	Y_1	J_0	K_0
0 0	1	1	d	1	1	d
0 1	0	d	1	0	d	1
1 0	1	1	d	1	1	d
1 1	0	d	1	0	d	1

Step 8: Derive the input equations for each flip-flop based as functions of the input and current state of all flip-flops.

At this point, we try to derive an expression that matches each column. Formal methods can be used, but generally are more trouble than they are worth. Here is this author's set of rules to match an expression to a given column.

1. If a column does not have a 0 in it, match it to the constant value 1. If a column does not have a 1 in it, match it to the constant value 0.

2. If the column has both 0's and 1's in it, try to match it to a single variable, which must be part of the present state. Only the 0's and 1's in a column must match the suggested function.

3. If every 0 and 1 in the column is a mismatch, match to the complement of a function.

4. If all the above fails, try for simple combinations of the present state.

The reader will note that there are two columns for each variable for which an equation is desired; one column for $X=0$ and one column for $X=1$. For example, consider the table for flip-flop 0, just above. If we work on a column-by column basis, we shall arrive at four equations.

One for J_0 when $X=0$,

one for K_0 when $X=0$,

one for J_0 when $X=1$, and

one for K_0 when $X=1$.

However, we need a single equation for J_0 and a single equation for K_0.

Registers and Counters

Registers

Registers are simply a set of flip-flops used to store a number of binary bits. A set of n flip-flops can store bits. 1-bit is a simple form of register. These registers consists of a 1-bit input (I), and enable signal (enb), and a 1-bit output (Q). When enb=1, the register loads the value of I on the rising clock edge. When enb=0, the register holds the current value.

| We create a module by name of **Reg1**. The register consists of an interface of 4 ports, clk, I, enb, Q. | ```
// 1-bit Register
module Reg1(clk, I
enb, Q);


endmodule
``` | *Reg1* |
| We next specify what type the ports are. In this example, clk, I, and enb are input ports. Port Q is an output port.<br><br>Notice in addition to declaring Q as an output, we must also declare Q as a register so we can assign values to Q. | ```
// 1-bit Register
module Reg1(clk, I enb, Q);


input clk;
input I, enb;
output Q;
reg Q;

endmodule
``` | *Reg1* |
| We then create a sensitivity list, which executes the code between the "begin" and "end" statements onv every rising edge of the clk. | ```
// 1-bit Register
module Reg1(clk, I
enb, Q);

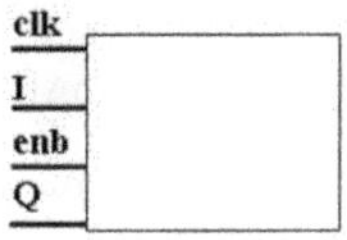

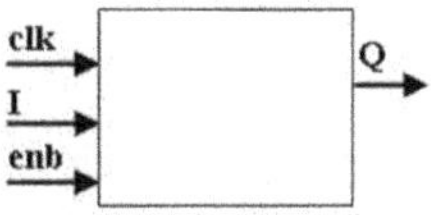

input clk;
input I, enb;
output Q;
reg Q;


always @ (posedge clk)
begin


end

endmodule
``` | *Reg1* |
| If enb=1 on the rising edge of the clk, then we store the value of input I. | ```
// 1-bit Register
module Reg1(clk, I
enb, Q);


input clk;
input I, enb;
output Q;
reg Q;


always @ (posedge clk)
begin
  if(enb == 1)
    Q <= I;
end

endmodule
``` | *Reg1* |

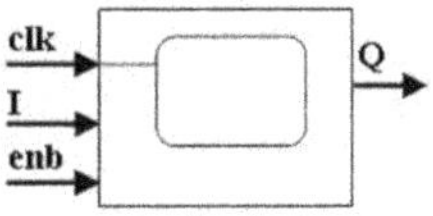

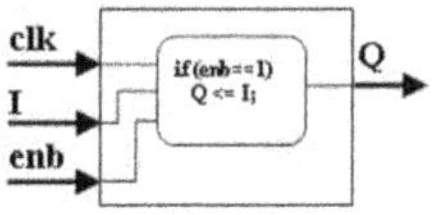

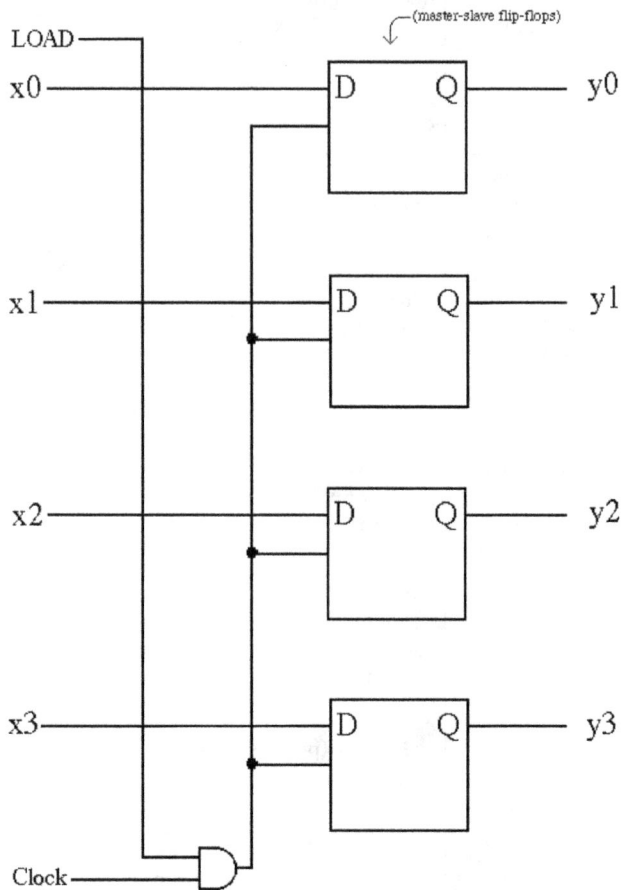

(master-slave flip-flops)

Description

A CPU contains very fast memory called registers. For example, a MIPS ISA stores 32-bit registers.

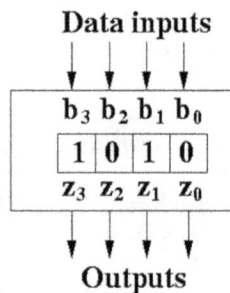

Data inputs

b_3	b_2	b_1	b_0
1	0	1	0
z_3	z_2	z_1	z_0

Outputs

In some sense, this is not really a black box, because we have some idea of what's in it. The black box shown above has four inputs. The register can do one of two things: it can parallel load the inputs, that is, it can read in all four inputs and store the four bits into the array. Each element of the array can store a single bit.

It can also choose to ignore the inputs, and the values of the arrays remain unchanged. In this case, we say the array is "holding" its values. The array's values are continuously being sent as output. Thus, if we could measure the output of the register (say, with a voltmeter), the four bits being output would correspond to the four bits currently being stored in the array.

This output is sent to the outside world where other devices can read the values if they want to. A register can have more than 4 bits stored in it. In fact, modern CPUs either store 32 or 64 bits in registers. In reality, one bit of a register is really a flip flop, which is a device that can store one bit. We will discuss flip flops at some later point.

Parallel Load Register

This register has a name. It's called a parallel load register. It supports two operations. The operation is controlled by a single control bit, called c.

c	Operation
0	Hold
1	Parallel Load

Like combinational logic circuits, the c refers to a control bit. Recall that control bits are used to determine which operation or function a device performs. Unlike combinational logic circuits, registers use clocks. Here's an important feature of registers (and sequential logic devices in general).

A device that is edge-triggered (which this register is) can only change values when there is a clock edge. In particular, we assume a register is positive-edge triggered, so it can only change values whenever a positive edge occurs.

Specifically, when the clock reaches a positive edge, the register detects this event. It then reads in the value of c, the control bit, and based on the value, either parallel loads (*i.e.*, sets $z_3 z_2 z_1 z_0 = x_3 x_2 x_1 x_0$) or it holds. If a clock is not at a positive edge, then the value of the registers is held and does not change.

Hold

You may wonder why we have a "hold" operation. After all, when there's no positive edge, the register seems to hold the value already. A positive edge occurs on a regular basis (just like the milkman appears on a daily basis), and it's not always the case that you want to parallel load every time this happens. You want to control when a register is loaded. Sometimes you want it to parallel load, sometimes to hold its value. Thus, the need for a control signal.

Registers are Memory

Registers store information over time, which makes them devices with memory. There are other kinds of circuits (combinational logic circuits and wires) which merely process inputs and produce outputs, but do not store any values.

Clock

When we study combinational logic, we won't use a clock there. Combinational logic circuits are made from AND gates, OR gates, *etc.* These circuits can be modeled using a graph. Such graphs are acyclic, *i.e.*, there are no cycles.

However, in general, it's useful to design circuits with feedback (*i.e.*, cycles). Circuits with feedback are easier to design if we can control how often the circuit is updated. Thus,the need for a clock. For now, just believe that it's important for a register to have a clock.

Let's look at a picture of a 4-bit register (Figure 8.33).

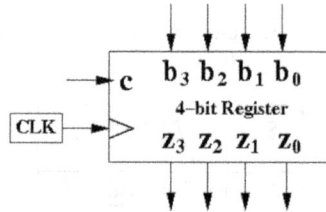

Figure 8.33: 4-Bit Register

A 4-bit register has the following inputs and outputs:

☆ 4 bits of data input $b_{3\text{-}0}$

☆ 1 control bit, c, which indicates whether to perform a parallel load or hold.

☆ A clock (which is an input), used for timing. The register will only perform the operation when there is a positive edge from the clock.

Parallel Load

Now suppose the circuit is attempting a parallel load. That is, $c == 1$. Assume the positive edge has not yet arrived, and the register looks like:

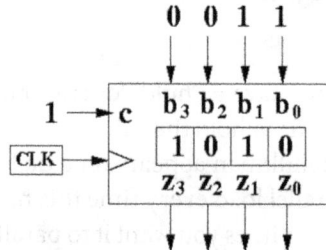

After a positive edge occurs, the 4 bits are *read in* from the input, in parallel. This overwrites the previous bits, and replaces it with the new bits. In this case, those bits are 0011.

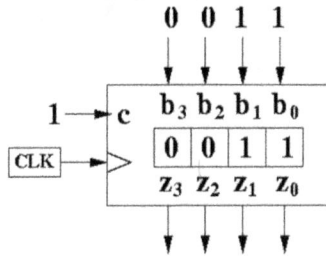

Once the bits are loaded, the outputs z_{3-0} is 0011.

Remember

- ☆ A k-bit register contains k flip flops.
- ☆ Each flip flop can store a single bit.
- ☆ A positive edge-triggered flip flop (thus, a register) can only change the value of the bit on a positive edge.
- ☆ When a positive edge occurs, a register can either hold its value (if $c == 0$) or parallel load (if $c == 1$).
- ☆ A parallel load means the bits are read in from the input bits ($b_{(k-1)-0}$) and stored in the k-bits within the register.
- ☆ A hold means the registers do not read in the bits, but maintains the current values of the bits.
- ☆ A register can only change its value at most once per positive edge.
- ☆ When the clock is not at a positive edge, the register maintains ("holds") its value.
- ☆ A k-bit register always outputs its values through $z_{(k-1)-0}$.

Counters

Counters can be classified into two broad categories according to the way they are clocked:

1. Asynchronous (Ripple) Counters - the first flip-flop is clocked by the external clock pulse, and then each successive flip-flop is clocked by the Q or Q' output of the previous flip-flop.

2. Synchronous Counters - all memory elements are simultaneously triggered by the same clock.

Binary Count Sequence

If we examine a four-bit binary count sequence from 0000 to 1111, a definite pattern will be evident in the "oscillations" of the bits between 0 and 1:

```
0 0 0 0
0 0 0 1
0 0 1 0
0 0 1 1
0 1 0 0
0 1 0 1
0 1 1 0
0 1 1 1
1 0 0 0
1 0 0 1
1 0 1 0
1 0 1 1
1 1 0 0
1 1 0 1
1 1 1 0
1 1 1 1
```

Note how the least significant bit (LSB) toggles between 0 and 1 for every step in the count sequence, while each succeeding bit toggles at one-half the frequency of the one before it. The most significant bit (MSB) only toggles once during the entire sixteen-step count sequence: at the transition between 7 (0111) and 8 (1000).

If we wanted to design a digital circuit to "count" in four-bit binary, all we would have to do is design a series of frequency divider circuits, each circuit dividing the frequency of a square-wave pulse by a factor of 2 (Figure 8.34):

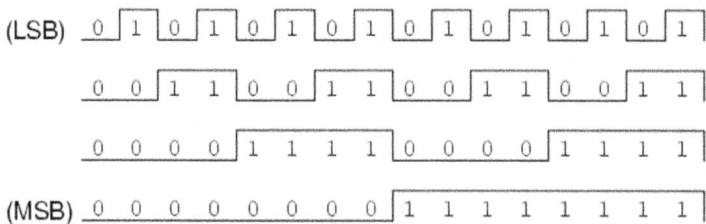

Figure 8.34: Frequency of a Square-Wave Pulse.

J-K flip-flops are ideally suited for this task because they have the ability to "toggle" their output state at the command of a clock pulse when both J and K inputs are made "high" (1) (Figure 8.35):

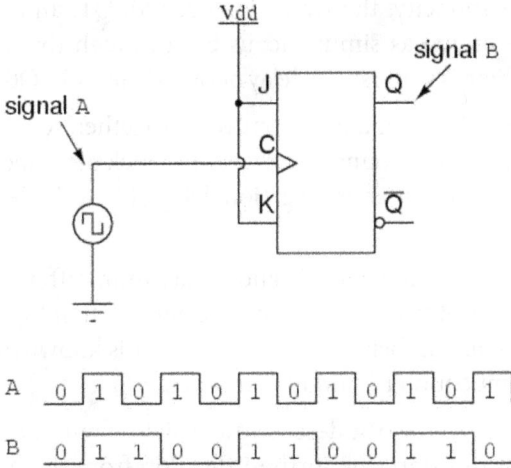

Figure 8.35: Signals of J-K Flip-Flops.

If we consider the two signals (A and B) in this circuit to represent two bits of a binary number, signal A being the LSB and signal B being the MSB, we see that the count sequence is backward: from 11 to 10 to 01 to 00 and back again to 11. Although it might not be counting in the direction we might have assumed, at least it counts.

Asynchronous (Ripple) Counters

A two-bit asynchronous counter is shown below. The external clock is connected to the clock input of the first flip-flop (FF0) only. So, FF0 changes state at the falling edge of each clock pulse, but FF1 changes only when triggered by the falling edge of the Q output of FF0. Because of the inherent propagation delay through a flip-flop, the transition of the input clock pulse and a transition of the Q output of FF0 can never occur at exactly the same time. Therefore, the flip-flops cannot be triggered simultaneously, producing an asynchronous operation.

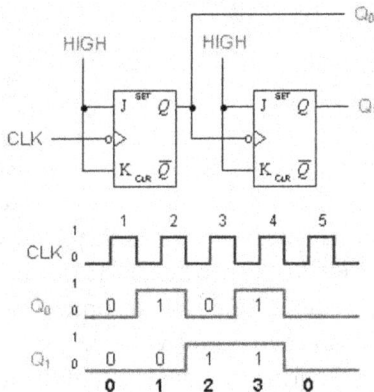

Note that for simplicity, the transitions of Q0, Q1, and CLK in the timing diagram above are shown as simultaneous even though this is an asynchronous counter. Actually, there is some small delay between the CLK, Q0 and Q1 transitions.

Usually, all the CLEAR inputs are connected together, so that a single pulse can clear all the flip-flops before counting starts. The clock pulse fed into FF0 is rippled through the other counters after propagation delays, like a ripple on the water, hence the name Ripple Counter.

The 2-bit ripple counter circuit above has four different states, each one corresponding to a count value. Similarly, a counter with *n* flip-flops can have *2 to the power n* states. The number of states in a counter is known as its mod (modulo) number. Thus a 2-bit counter is amod-4 counter.

A mod-n counter may also be described as a divide-by-*n* counter. This is because the most significant flip-flop (the furthest flip-flop from the original clock pulse) produces one pulse for every *n* pulses at the clock input of the least significant flip-flop (the one triggers by the clock pulse). Thus, the above counter is an example of a divide-by-4 counter.

The following is a three-bit asynchronous binary counter and its timing diagram for one cycle. It works exactly the same way as a two-bit asynchronous binary counter mentioned above, except it has eight states due to the third flip-flop.

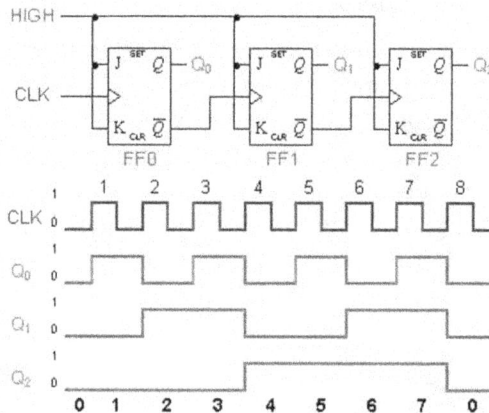

Since we know that binary count sequences follow a pattern of octave (factor of 2) frequency division and that J-K flip-flop multivibrators set up for the "toggle" mode are capable of performing this type of frequency division, we can envision a circuit made up of several J-K flip-flops, cascaded to produce four bits of output. The main problem facing us is to determine *how* to connect these flip-flops together so that they toggle at the right times to produce the proper binary sequence. Examine the following binary count sequence, paying attention to patterns preceding the "toggling" of a bit between 0 and 1:

```
0 0 0 0
0 0 0 1
0 0 1 0
0 0 1 1
0 1 0 0
0 1 0 1
0 1 1 0
0 1 1 1
1 0 0 0
1 0 0 1
1 0 1 0
1 0 1 1
1 1 0 0
1 1 0 1
1 1 1 0
1 1 1 1
```

Note that each bit in this four-bit sequence toggles when the bit before it (the bit having a lesser significance, or place-weight), toggles in a particular direction: from 1 to 0. Small arrows indicate those points in the sequence where a bit toggles, the head of the arrow pointing to the previous bit transitioning from a "high" (1) state to a "low" (0) state:

```
0 0 0 0
0 0 0 1
       →
0 0 1 0
0 0 1 1
      →→
0 1 0 0
0 1 0 1
       →
0 1 1 0
0 1 1 1
    →→→→
1 0 0 0
1 0 0 1
       →
1 0 1 0
1 0 1 1
      →→
1 1 0 0
1 1 0 1
       →
1 1 1 0
1 1 1 1
```

Starting with four J-K flip-flops connected in such a way to always be in the "toggle" mode, we need to determine how to connect the clock inputs in such a way so that each succeeding bit toggles when the bit before it transitions from 1 to 0. The Q outputs of each flip-flop will serve as the respective binary bits of the final, four-bit count (Figure 8.36):

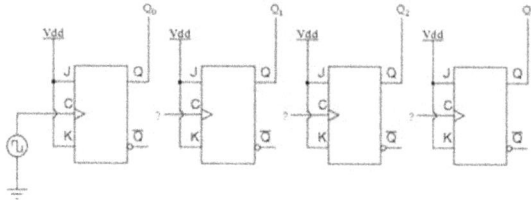

Figure 8.36: Circuit of Four J-K Flip-Flops.

If we used flip-flops with negative-edge triggering (bubble symbols on the clock inputs), we could simply connect the clock input of each flip-flop to the Q output of the flip-flop before it, so that when the bit before it changes from a 1 to a 0, the "falling edge" of that signal would "clock" the next flip-flop to toggle the next bit:

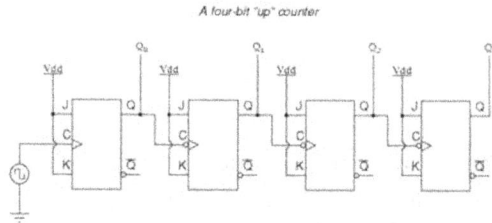

This circuit would yield the following output waveforms, when "clocked" by a repetitive source of pulses from an oscillator (Figure 8.37):

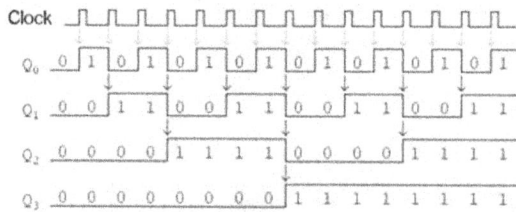

Figure 8.37: Output Waveforms.

The first flip-flop (the one with the Q_0 output), has a positive-edge triggered clock input, so it toggles with each rising edge of the clock signal. Notice how the clock signal in this example has a duty cycle less than 50%. I've shown the signal in this manner for the purpose of demonstrating how the clock signal need not be symmetrical to obtain reliable, "clean" output bits in our four-bit binary sequence. In the very first flip-flop circuit shown in this chapter, I used the clock signal itself as one of the output bits. This is a bad practice in counter design, though, because it necessitates the use of a square wave signal with a 50% duty cycle ("high" time =

"low" time) in order to obtain a count sequence where each and every step pauses for the same amount of time. Using one J-K flip-flop for each output bit, however, relieves us of the necessity of having a symmetrical clock signal, allowing the use of practically any variety of high/low waveform to increment the count sequence.

As indicated by all the other arrows in the pulse diagram, each succeeding output bit is toggled by the action of the preceding bit transitioning from "high" (1) to "low" (0). This is the pattern necessary to generate an "up" count sequence.

A less obvious solution for generating an "up" sequence using positive-edge triggered flip-flops is to "clock" each flip-flop using the Q' output of the preceding flip-flop rather than the Q output. Since the Q' output will always be the exact opposite state of the Q output on a J-K flip-flop (no invalid states with this type of flip-flop), a high-to-low transition on the Q output will be accompanied by a low-to-high transition on the Q' output. In other words, each time the Q output of a flip-flop transition from 1 to 0, the Q' output of the same flip-flop will transition from 0 to 1, providing the positive-going clock pulse we would need to toggle a positive-edge triggered flip-flop at the right moment:

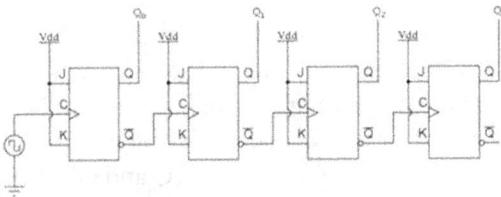

A different way of making a four-bit "up" counter

One way we could expand the capabilities of either of these two counter circuits is to regard the Q' outputs as another set of four binary bits. If we examine the pulse diagram for such a circuit, we see that the Q' outputs generate a *down*-counting sequence, while the Q outputs generate an *up*-counting sequence (Figure 8.38):

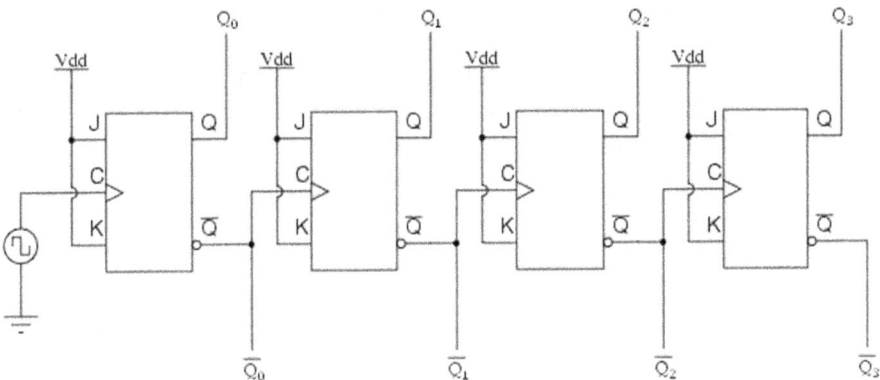

A simultaneous "up" and "down" counter

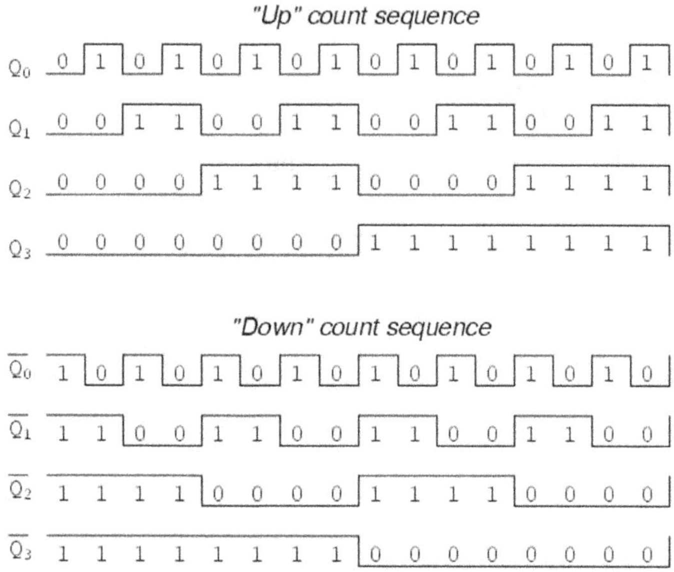

"Up" count sequence

Q_0 0 1 0 1 0 1 0 1 0 1 0 1 0 1 0 1

Q_1 0 0 1 1 0 0 1 1 0 0 1 1 0 0 1 1

Q_2 0 0 0 0 1 1 1 1 0 0 0 0 1 1 1 1

Q_3 0 0 0 0 0 0 0 0 1 1 1 1 1 1 1 1

"Down" count sequence

$\overline{Q_0}$ 1 0 1 0 1 0 1 0 1 0 1 0 1 0 1 0

$\overline{Q_1}$ 1 1 0 0 1 1 0 0 1 1 0 0 1 1 0 0

$\overline{Q_2}$ 1 1 1 1 0 0 0 0 1 1 1 1 0 0 0 0

$\overline{Q_3}$ 1 1 1 1 1 1 1 1 0 0 0 0 0 0 0 0

Figure 8.38: UP/Down Count Sequence.

When the Q output of a flip-flop transition from 1 to 0, it commands the next flip-flop to toggle. If the next flip-flop toggle is a transition from 1 to 0, it will command the flip-flop after it to toggle as well, and so on. However, since there is always some small amount of propagation delay between the command to toggle (the clock pulse) and the actual toggle response (Q and Q' outputs changing states), any subsequent flip-flops to be toggled will toggle sometime *after* the first flip-flop has toggled. Thus, when multiple bits toggle in a binary count sequence, they will not all toggles at exactly the same time:

Asynchronous Decade Counters

The binary counters previously introduced have two to the power *n* states. But counters with states less than this number are also possible. They are designed to have the number of states in their sequences, which are called truncated sequences. These sequences are achieved by forcing the counter to recycle before going through all of its normal states. A common modulus for counters with truncated sequences is ten. A counter with ten states in its sequence is called a *decade counter*. The circuit below is an implementation of a decade counter.

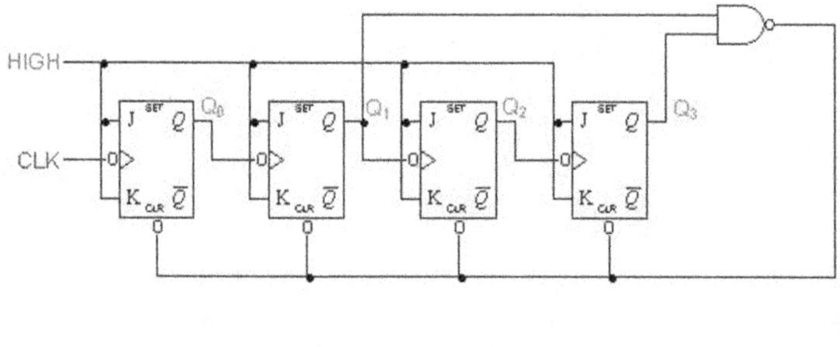

Once the counter counts to ten (1010), all the flip-flops are being cleared. Notice that only Q1 and Q3 are used to decode the count of ten. This is called partial decoding, as none of the other states (zero to nine) have both Q1 and Q3 HIGH at the same time. The sequence of the decade counter is shown in the table below:

Clock Pulse	Q3	Q2	Q1	Q0
0	0	0	0	0
1	0	0	0	1
2	0	0	1	0
3	0	0	1	1
4	0	1	0	0
5	0	1	0	1
6	0	1	1	0
7	0	1	1	1
8	1	0	0	0
9	1	0	0	1

Asynchronous Up-Down Counters

In certain applications, a counter must be able to count both up and down. The circuit below is a 3-bit up-down counter. It counts up or down depending on the status of the control signals UP and DOWN. When the UP input is at 1 and the DOWN input is at 0, the NAND network between FF0 and FF1 will gate the non-inverted output (Q) of FF0 into the clock input of FF1. Similarly, Q of FF1 will be gated through the other NAND network into the clock input of FF2. Thus the counter will count up.

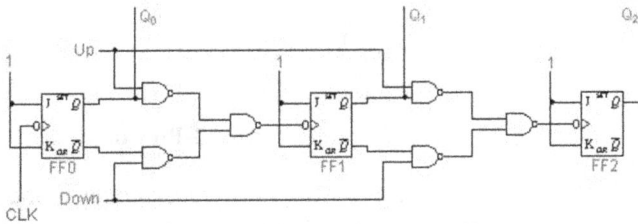

When the control input UP is at 0 and DOWN is at 1, the inverted outputs of FF0 and FF1 are gated into the clock inputs of FF1 and FF2 respectively. If the flip-flops are initially reset to 0's, then the counter will go through the following sequence as input pulses are applied.

FF2	FF1	FF0
0	0	0
1	1	1
1	1	0
1	0	1
1	0	0
0	1	1
0	1	0
0	0	1

Notice that an asynchronous up-down counter is slower than an up counter or a down counter because of the additional propagation delay introduced by the NAND networks.

Synchronous Counters

In *synchronous counters*, the clock inputs of all the flip-flops are connected together and are triggered by the input pulses. Thus, all the flip-flops change state simultaneously (in parallel). The circuit below is a 3-bit synchronous counter. The J and K inputs of FF0 are connected to HIGH. FF1 has its J and K inputs connected to the output of FF0, and the J and K inputs of FF2 are connected to the output of an AND gate that is fed by the outputs of FF0 and FF1.

The count sequence for the 3-bit counter is shown in Figure 8.39. The most important advantage of synchronous counters is that there is no cumulative time delay because all flip-flops are triggered in parallel. Thus, the maximum operating frequency for this counter will be significantly higher than for the corresponding ripple counter.

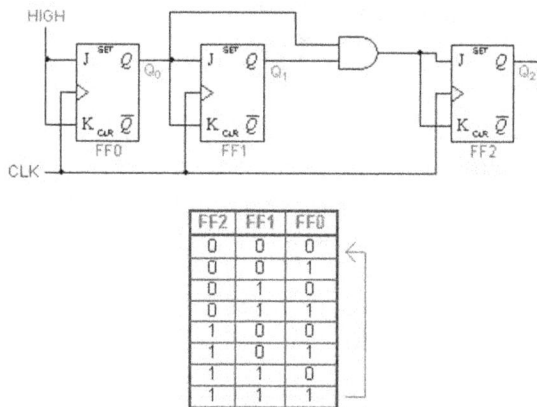

FF2	FF1	FF0
0	0	0
0	0	1
0	1	0
0	1	1
1	0	0
1	0	1
1	1	0
1	1	1

Figure 8.39: Count Sequence for the 3-Bit Counter.

A *synchronous counter*, in contrast to an *asynchronous counter*, is one whose output bits change state simultaneously, with no ripple. The only way we can build such a counter circuit from J-K flip-flops is to connect all the clock inputs together so that each and every flip-flop receives the exact same clock pulse at the exact same time:

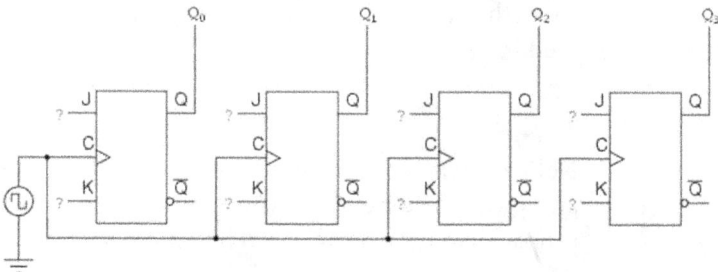

Now, the question is, what do we do with the J and K input? We know that we still have to maintain the same divide-by-two frequency pattern in order to count in a binary sequence and that this pattern is best achieved utilizing the "toggle" mode of the flip-flop, so the fact that the J and K inputs must both be (at times) "high" is clear. However, if we simply connect all the J and K inputs to the positive rail of the power supply as we did in the asynchronous circuit, this would clearly not work because all the flip-flops would toggle at the same time: with each and every clock pulse!

This circuit will not function as a counter!

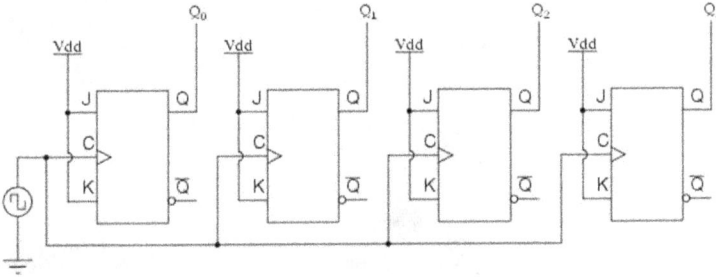

Let's examine the four-bit binary counting sequence again, and see if there are any other patterns that predict the toggling of a bit. Asynchronous counter circuit design is based on the fact that each bit toggle happens at the same time that the preceding bit toggles from a "high" to a "low" (from 1 to 0). Since we cannot clock the toggling of a bit based on the toggling of a previous bit in a synchronous counter circuit (to do so would create a ripple effect) we must find some other pattern in the counting sequence that can be used to trigger a bit toggle:

Examining the four-bit binary count sequence, another predictive pattern can be seen. Notice that just before a bit toggles, all preceding bits are "high:"

```
0 0 0 0
0 0 0 1
0 0 1 0
0 0 1 1
0 1 0 0
0 1 0 1
0 1 1 0
0 1 1 1
1 0 0 0
1 0 0 1
1 0 1 0
1 0 1 1
1 1 0 0
1 1 0 1
1 1 1 0
1 1 1 1
```

This pattern is also something we can exploit in designing a counter circuit. If we enable each J-K flip-flop to toggle based on whether or not all preceding flip-flop outputs (Q) are "high," we can obtain the same counting sequence as the asynchronous circuit without the ripple effect since each flip-flop in this circuit will be clocked at exactly the same time:

A four-bit synchronous "up" counter

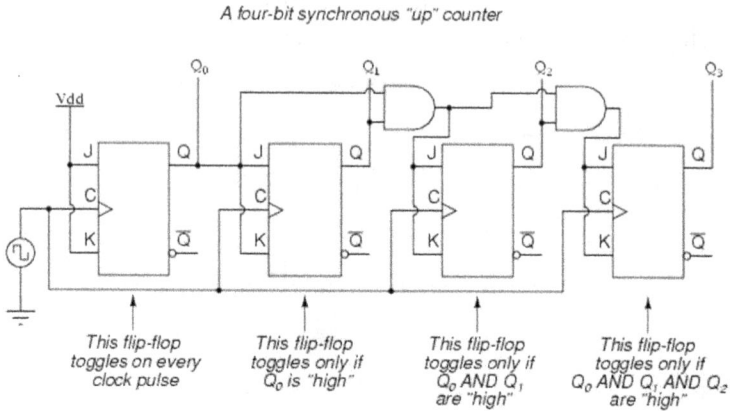

This flip-flop toggles on every clock pulse

This flip-flop toggles only if Q_0 is "high"

This flip-flop toggles only if Q_0 AND Q_1 are "high"

This flip-flop toggles only if Q_0 AND Q_1 AND Q_2 are "high"

The result is a four-bit *synchronous* "up" counter. Each of the higher-order flip-flops is made ready to toggle (both J and K inputs "high") if the Q outputs of all previous flip-flops are "high." Otherwise, the J and K inputs for that flip-flop will both be "low," placing it into the "latch" mode where it will maintain its present output state at the next clock pulse. Since the first (LSB) flip-flop needs to toggle at every clock pulse, its J and K inputs are connected to V_{cc} or V_{dd}, where they will be "high" all the time. The next flip-flop need only "recognize" that the first flip-flop's Q output is high to be made ready to toggle, so no AND gate is needed. However, the remaining flip-flops should be made ready to toggle only when *all* lower-order output bits are "high," thus the need for AND gates.

To make a synchronous "down" counter, we need to build the circuit to recognize the appropriate bit patterns predicting each toggle state while counting down. Not surprisingly, when we examine the four-bit binary count sequence, we see that all preceding bits are "low" prior to a toggle (following the sequence from bottom to top):

0 0 0 0
0 0 0 1
0 0 1 0
0 0 1 1
0 1 0 0
0 1 0 1
0 1 1 0
0 1 1 1
1 0 0 0
1 0 0 1
1 0 1 0
1 0 1 1
1 1 0 0
1 1 0 1
1 1 1 0
1 1 1 1

Since each J-K flip-flop comes equipped with a Q' output as well as a Q output, we can use the Q' outputs to enable the toggle mode on each succeeding flip-flop, being that each Q' will be "high" every time that the respective Q is "low:"

A four-bit synchronous "down" counter

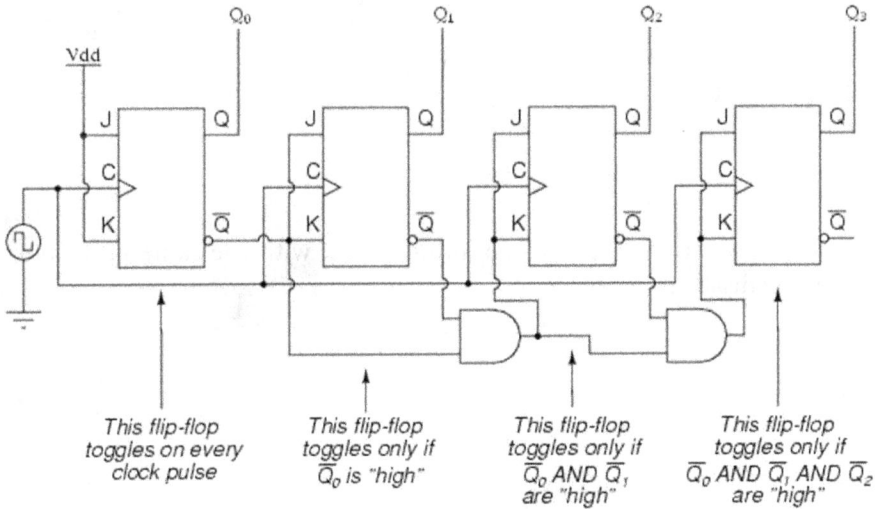

| This flip-flop toggles on every clock pulse | This flip-flop toggles only if $\overline{Q}_0$ is "high" | This flip-flop toggles only if $\overline{Q}_0$ AND $\overline{Q}_1$ are "high" | This flip-flop toggles only if $\overline{Q}_0$ AND $\overline{Q}_1$ AND $\overline{Q}_2$ are "high" |

Taking this idea one step further, we can build a counter circuit with selectable between "up" and "down" count modes by having dual lines of AND gates detecting the appropriate bit conditions for an "up" and a "down" counting sequence, respectively, then use OR gates to combine the AND gate outputs to the J and K inputs of each succeeding flip-flop:

A four-bit synchronous "up/down" counter

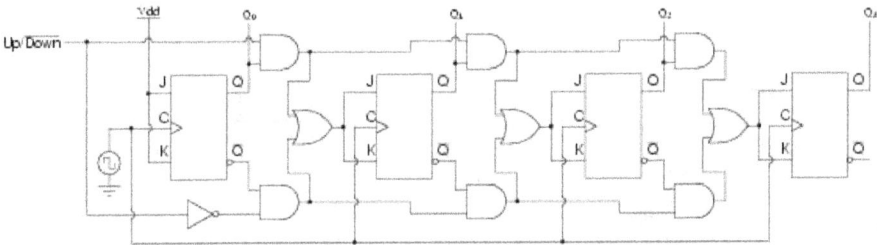

This circuit isn't as complex as it might first appear. The Up/Down control input line simply enables either the upper string or lower string of AND gates to pass the Q/Q' outputs to the succeeding stages of flip-flops. If the Up/Down control line is "high," the top AND gates become enabled, and the circuit functions exactly the same as the first ("up") synchronous counter circuit shown in this section. If the Up/Down control line is made "low," the bottom AND gates become enabled, and the circuit functions identically to the second ("down" counter) circuit shown in this section.

To illustrate, here is a diagram showing the circuit in the "up" counting mode (all disabled circuitry shown in grey rather than black):

Counter in "up" counting mode

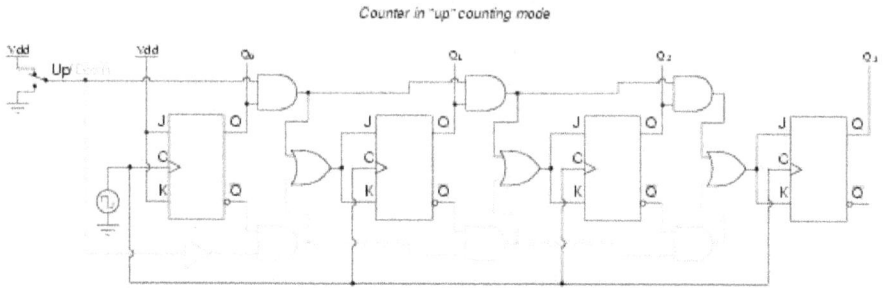

Here, shown in the "down" counting mode, with the same grey coloring representing disabled circuitry:

Counter in "down" counting mode

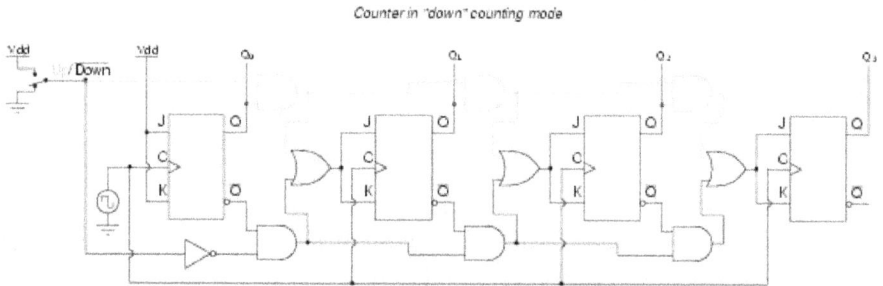

Up/down counter circuits are very useful devices. A common application is in machine motion control, where devices called *rotary shaft encoders* convert mechanical rotation into a series of electrical pulses, these pulses "clocking" a counter circuit to track total motion:

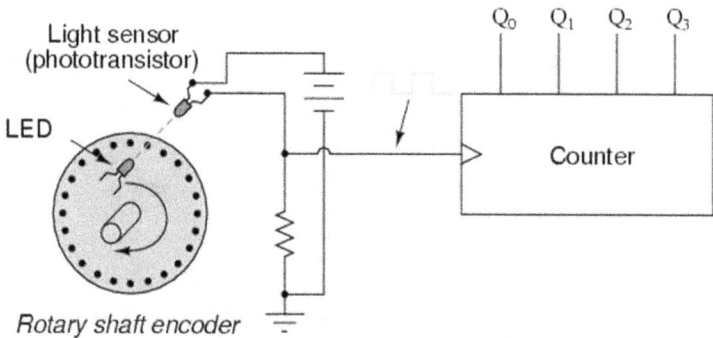

As the machine moves, it turns the encoder shaft, making and breaking the light beam between LED and phototransistor, thereby generating clock pulses to increment the counter circuit. Thus, the counter integrates or accumulates, total motion of the shaft, serving as an electronic indication of how far the machine has moved. If all we care about is tracking total motion, and do not care to account for changes in the *direction* of motion, this arrangement will suffice. However, if we wish the counter to *increment* with one direction of motion and *decrement* with the reverse direction of motion, we must use an up/down counter, and an encoder/decoding circuit having the ability to discriminate between different directions.

If we re-design the encoder to have two sets of LED/phototransistor pairs, those pairs aligned such that their square-wave output signals are 90° out of phase with each other, we have what is known as a *quadrature output* encoder (the word "quadrature" simply refers to a 90° angular separation). A phase detection circuit may be made from a D-type flip-flop, to distinguish a clockwise pulse sequence from a counter-clockwise pulse sequence:

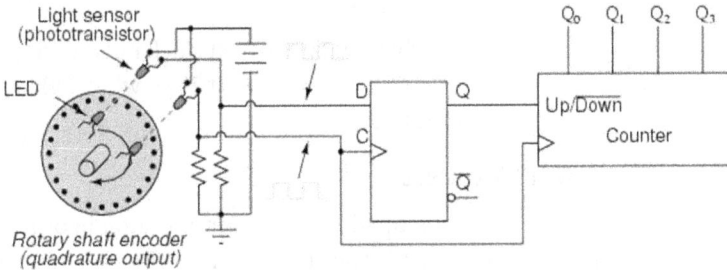

When the encoder rotates clockwise, the "D" input signal square-wave will lead the "C" input square-wave, meaning that the "D" input will already be "high" when the "C" transitions from "low" to "high," thus *setting* the D-type flip-flop (making the Q output "high") with every clock pulse. A "high" Q output places the counter into the "Up" count mode, and any clock pulses received by the clock from the encoder (from either LED) will increment it. Conversely, when the encoder reverses rotation, the "D" input will lag behind the "C" input waveform, meaning that it will be "low" when the "C" waveform transitions from "low" to "high," forcing the D-type flip-flop into the *reset* state (making the Q output "low") with every clock pulse. This "low" signal commands the counter circuit to decrement with every clock pulse from the encoder.

Synchronous Decade Counters

Similar to an asynchronous decade counter, a *synchronous decade counter* counts from 0 to 9 and then recycles to 0 again. This is done by forcing the 1010 state back to the 0000 state. This so called truncated sequence can be constructed by the following circuit.

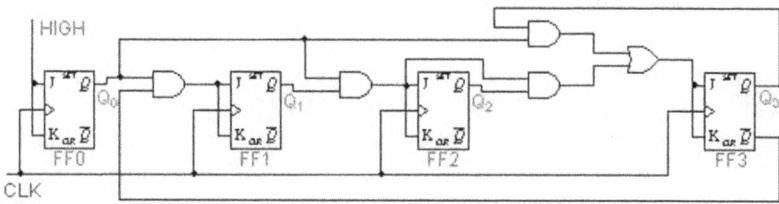

From the sequence on the left, we notice that:

Clock Pulse	Q3	Q2	Q1	Q0
0	0	0	0	0
1	0	0	0	1
2	0	0	1	0
3	0	0	1	1
4	0	1	0	0
5	0	1	0	1
6	0	1	1	0
7	0	1	1	1
8	1	0	0	0
9	1	0	0	1

☆ Q0 toggles on each clock pulse.

☆ Q1 changes on the next clock pulse each time Q0=1 and Q3=0.

☆ Q2 changes on the next clock pulse each time Q0=Q1=1.

☆ Q3 changes on the next clock pulse each time Q0=1, Q1=1 and Q2=1 (count 7), or when Q0=1 and Q3=1 (count 9).

Synchronous Up-Down Counters

A circuit of a 3-bit synchronous up-down counter and a table of its sequence are shown below. Similar to an asynchronous up-down counter, a synchronous up-down counter also has an up-down control input. It is used to control the direction of the counter through a certain sequence.

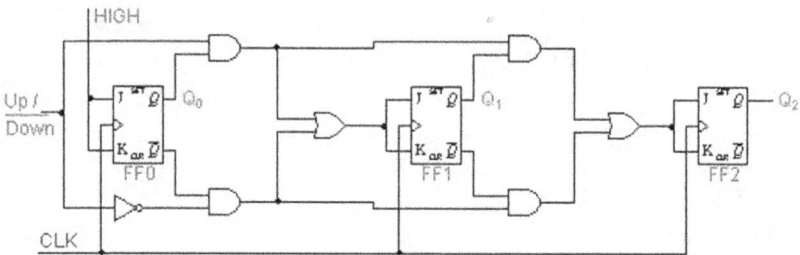

Up	Q2	Q1	Q0	Down
	0	0	0	
	0	0	1	
	0	1	0	
	0	1	1	
	1	0	0	
	1	0	1	
	1	1	0	
	1	1	1	

An examination of the sequence table shows:

☆ In both the UP and DOWN sequences, Q0 toggles on each clock pulse.

☆ In the UP sequence, Q1 changes state on the next clock pulse when Q0=1.

☆ In the DOWN sequence, Q1 changes state on the next clock pulse when Q0=0.

☆ In the UP sequence, Q2 changes state on the next clock pulse when Q0=Q1=1.

☆ In the DOWN sequence, Q2 changes state on the next clock pulse when Q0=Q1=0.

Shift Registers

Shift registers are a type of sequential logic circuit, mainly for storage of digital data. They are a group of flip-flops connected in a chain so that the output from one flip-flop becomes the input of the next flip-flop. Most of the registers possess no characteristic internal sequence of states. All the flip-flops are driven by a common clock, and all are set or reset simultaneously.

Description

Shift registers, like counters, are a form of *sequential logic*. Sequential logic, unlike combinational logic, is not only affected by the present inputs but also, by the prior history. In other words, sequential logic remembers past events.

Shift registers produce a discrete delay of a digital signal or waveform. A waveform synchronized to a *clock*, a repeating square wave, is delayed by "**n**" discrete clock times, where "**n**" is the number of shift register stages. Thus, a four- stage shift register delays "data in" by four clocks to "data out". The stages in a shift register are *delay stages*, typically type "**D**" Flip-Flops or type "**JK**" Flip-flops.

Formerly, very long (several hundred stages) shift registers served as digital memory. This obsolete application is reminiscent of the acoustic mercury delay lines used as early computer memory.

Serial data transmission, over a distance of meters to kilometers, uses shift registers to convert parallel data to serial form. Serial data communications replace many slow parallel data wires with a single serial high- speed circuit.

Serial data over shorter distances of tens of centimeters uses shift registers to get data into and out of microprocessors. Numerous peripherals, including the analog to digital converters, digital to analog converters, display drivers, and memory, use shift registers to reduce the amount of wiring in circuit boards.

Some specialized counter circuits actually use shift registers to generate repeating waveforms. Longer shift registers, with the help of feedback, generate patterns so long that they look like random noise, *pseudo-noise*.

Basic shift registers are classified by structure according to the following types:

☆ Serial-in/serial-out

☆ Parallel-in/serial-out

☆ Serial-in/parallel-out

☆ Universal parallel-in/parallel-out

☆ Ring counter

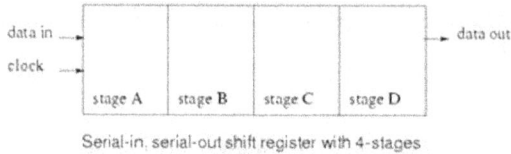

Serial-in, serial-out shift register with 4-stages

Above we show a block diagram of a serial-in/serial-out shift register, which is 4-stages long. Data at the input will be delayed by four clock periods from the input to the output of the shift register.

Data at "data in", above, will be present at the Stage **A** output after the first clock pulse. After the second pulse stage **A** data is transferred to stage **B** output, and "data in" is transferred to stage **A** output. After the third clock, stage C is replaced by stage **B**; stage **B** is replaced by stage **A**; and stage A is replaced by "data in". After the fourth clock, the data originally present at "data in" is at stage **D**, "output". The "first in" data is "first out" as it is shifted from "data in" to "data out".

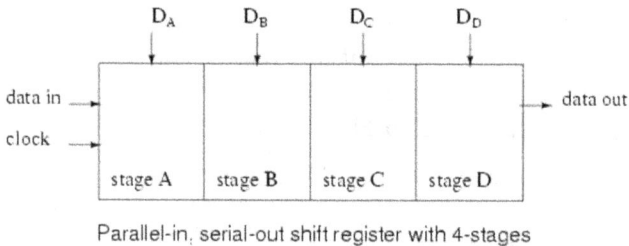

Parallel-in, serial-out shift register with 4-stages

Data is loaded into all stages at once of a parallel-in/serial-out shift register. The data is then shifted out via "data out" by clock pulses. Since a 4- stage shift register is shown above, four clock pulses are required to shift out all of the data. In the diagram above, stage **D** data will be present at the "data out" up until the first clock pulse; stage **C** data will be present at "data out" between the first clock and the second clock pulse; stage **B** data will be present between the second clock and the third clock; and stage **A** data will be present between the third and the fourth clock. After the fourth clock pulse and thereafter, successive bits of "data in" should appear at "data out" of the shift register after a delay of four clock pulses.

If four switches were connected to D_A through D_D, the status could be read into a microprocessor using only one data pin and a clock pin. Since adding more switches would require no additional pins, this approach looks attractive for many inputs.

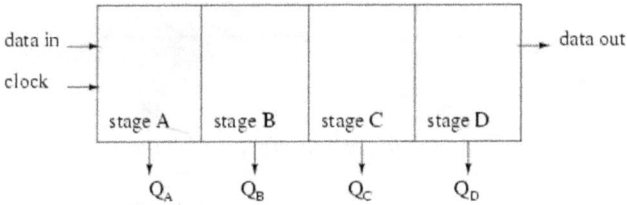

Serial-in, parallel-out shift register with 4-stages

Above, four data bits will be shifted in from "data in" by four clock pulses and be available at Q_A through Q_D for driving external circuitry such as LEDs, lamps, relay drivers, and horns.

After the first clock, the data at "data in" appears at Q_A. After the second clock, The old Q_A data appears at Q_B; Q_A receives next data from "data in". After the third clock, Q_B data is at Q_C. After the fourth clock, Q_C data is at Q_D. This stage contains the data first present at "data in". The shift register should now contain four data bits.

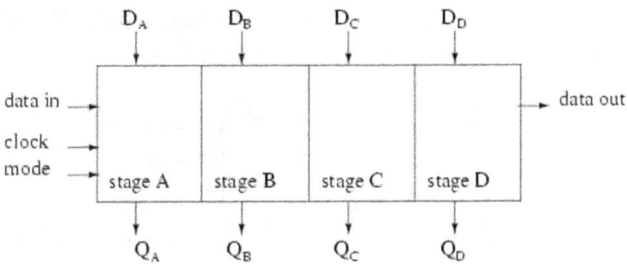

Parallel-in, parallel-out shift register with 4-stages

A parallel-in/parallel-out shift register combines the function of the parallel-in, serial-out shift register with the function of the serial-in, parallel-out shift register to yields the universal shift register. The "do anything" shifter comes at a price– the increased number of I/O (Input/Output) pins may reduce the number of stages which can be packaged.

Data presented at D_A through D_D is parallel loaded into the registers. This data at Q_A through Q_D may be shifted by the number of pulses presented at the clock input. The shifted data is available at Q_A through Q_D. The "mode" input, which may be more than one input, controls parallel loading of data from D_A through D_D, shifting of data, and the direction of shifting. There are shift registers which will shift data to either left or right.

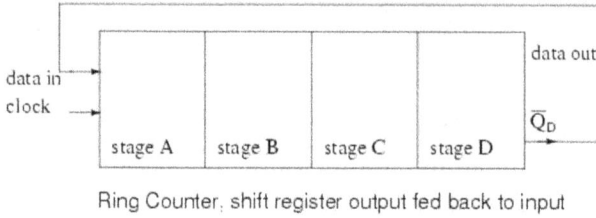

Ring Counter, shift register output fed back to input

If the serial output of a shift register is connected to the serial input, data can be perpetually shifted around the ring as long as clock pulses are present. If the output is inverted before being fed back, as shown above, we do not have to worry about loading the initial data into the "ring counter".

Serial In - Serial Out Shift Registers

A basic four-bit shift register can be constructed using four D flip-flops, as shown below. The operation of the circuit is as follows. The register is first cleared, forcing all four outputs to zero. The input data is then applied sequentially to the D input of the first flip-flop on the left (FF0). During each clock pulse, one bit is transmitted from left to right. Assume a data word to be 1001. The least significant bit of the data has to be shifted through the register from FF0 to FF3.

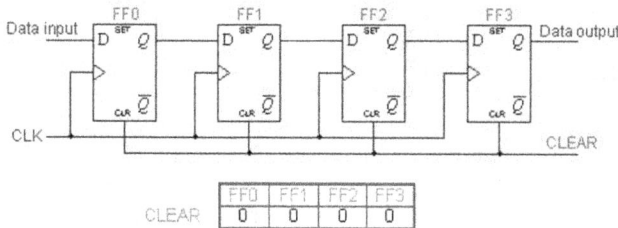

In order to get the data out of the register, they must be shifted out serially. This can be done destructively or non-destructively. For destructive readout, the original data is lost and at the end of the read cycle, all flip-flops are reset to zero.

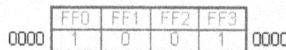

To avoid the loss of data, an arrangement for a non-destructive reading can be done by adding two AND gates, an OR gate and an inverter to the system. The construction of this circuit is shown below.

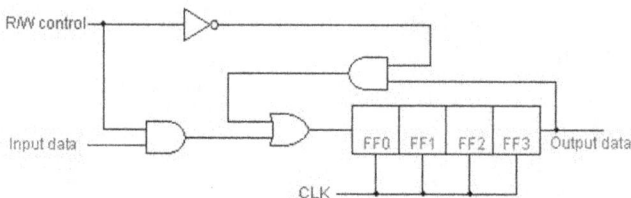

The data is loaded to the register when the control line is HIGH (ie WRITE). The data can be shifted out of the register when the control line is LOW (ie READ).

WRITE	FF0	FF1	FF2	FF3
1001	0	0	0	0

Serial-in, serial-out shift registers delay data by one clock time for each stage. They will store a bit of data for each register. A serial-in, serial-out shift register may be one to 64 bits in length, longer if registers or packages are cascaded.

Below is a single stage shift register receiving data which is not synchronized to the register clock. The "data in" at the **D** pin of the type **D FF** (Flip-Flop) does not change levels when the clock changes from low to high. We may want to synchronize the data to a system- wide clock in a circuit board to improve the reliability of a digital logic circuit.

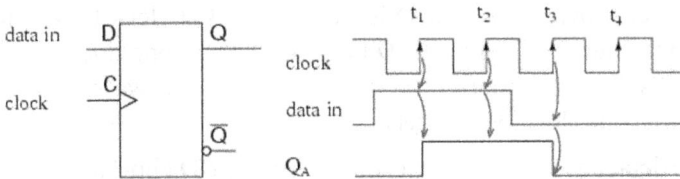

Data present at clock time is transfered from D to Q.

The obvious point illustrated above is that whatever "data in" is present at the **D** pin of a type **D** FF is transferred from D to output Q at clock time. Since our example shift register uses positive edge sensitive storage elements, the output **Q** follows the **D** input when the clock transitions from low to high as shown by the up arrows on the diagram above. There is no doubt what logic level is present at clock time because the data is stable well before and after the clock edge. This is seldom the case in multi-stage shift registers. But, this was an easy example to start with. We are only concerned with the positive, low to high, clock edge. The falling edge can be ignored. It is very easy to see **Q** follow **D** at clock time above. Compare this to the diagram below where the "data in" appears to change with the positive clock edge.

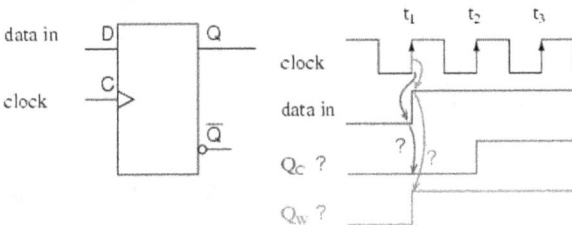

Does the clock t_1 see a 0 or a 1 at data in at D? Which output is correct, Q_C or Q_W?

Since "data in" appears to changes at clock time t_1 above, what does the type **D** FF see at clock time? The short over simplified answer is that it sees the data that was present at **D** prior to the clock. That is what is transferred to **Q** at clock time t_1. The correct waveform is Q_C. At t_1 Q goes to a zero if it is not already zero. The **D** register does not see a one until time t_2, at which time Q goes high.

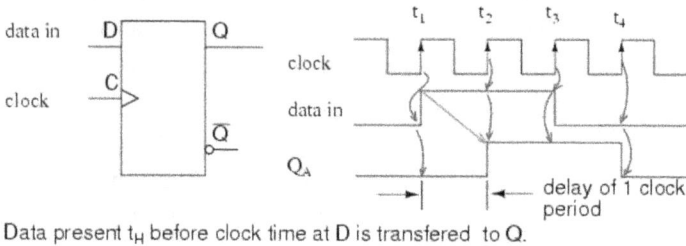

Data present t_H before clock time at D is transfered to Q.

Since data, above, present at **D** is clocked to **Q** at clock time, and **Q** cannot change until the next clock time, the **D** FF delays data by one clock period, provided that the data is already synchronized to the clock. The Q_A waveform is the same as "data in" with a one clock period delay.

A more detailed look at what the input of the type **D** Flip-Flop sees at clock time follows. Refer to the figure below. Since "data in" appears to changes at clock time (above), we need further information to determine what the **D** FF sees. If the "data in" is from another shift register stage, another same type **D** FF, we can draw some conclusions based on *data sheet* information. Manufacturers of digital logic make available information about their parts in data sheets, formerly only available in a collection called a *data book*. Databooks are still available; though, the manufacturer's website is the modern source.

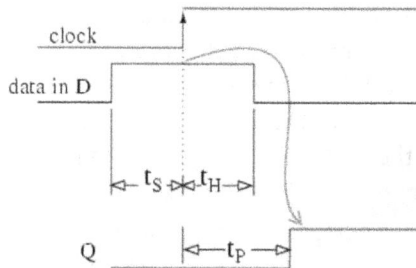

Data must be present (t_S) before the clock and after(t_H) the clock. Data is delayed from D to Q by propagation delay (t_P)

The following data was extracted from the CD4006b data sheet for operation at $5V_{DC}$, which serves as an example to illustrate timing.

- ☆ t_S=100ns
- ☆ t_H=60ns
- ☆ t_P=200-400ns typ/max

t_S is the *setup time*, the time data must be present before clock time. In this case, data must be present at **D** 100ns prior to the clock. Furthermore, the data must be held for *hold time* t_H=60ns after clock time. These two conditions must be met to reliably clock data from **D** to **Q** of the Flip-Flop.

There is no problem meeting the setup time of 60ns as the data at **D** has been there for the whole previous clock period if it comes from another shift register stage. For example, at a clock frequency of 1 Mhz, the clock period is 1000 µs, plenty of time. Data will actually be present for 1000µs prior to the clock, which is much greater than the minimum required t_S of 60ns.

The hold time t_H=60ns is met because D connected to Q of another stage cannot change any faster than the propagation delay of the previous stage t_p=200ns. Hold time is met as long as the propagation delay of the previous **D** FF is greater than the hold time. Data at **D** driven by another stage **Q** will not change any faster than 200ns for the CD4006b.

To summarize, output **Q** follows input D at nearly clock time if Flip-Flops are cascaded into a multi-stage shift register.

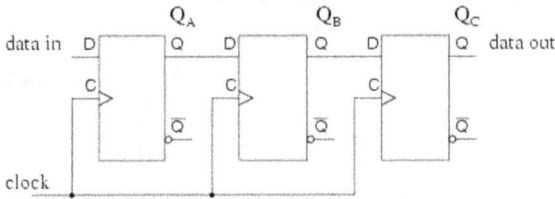

Serial-in, serial-out shift register using type "D" storage elements

Three type **D** Flip-Flops are cascaded Q to D and the clocks paralleled to form a three- stage shift register above.

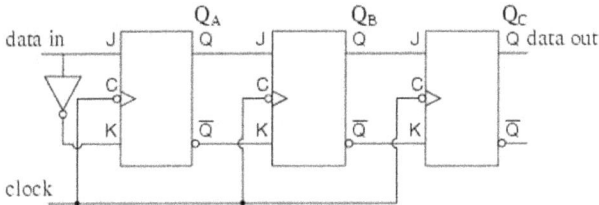

Serial-in, serial-out shift register using type "JK" storage elements

Type **JK** FFs cascaded Q to J, Q' to K with clocks in parallel to yield an alternate form of the shift register above.

A serial-in/serial-out shift register has a clock input, a data input, and a data output from the last stage. In general, the other stage outputs are not available Otherwise, it would be a serial-in, parallel-out shift register.

The waveforms below are applicable to either one of the preceding two versions of the serial-in, serial-out shift register. The three pairs of arrows show that a three-stage shift register temporarily stores 3-bits of data and delays it by three clock periods from input to output.

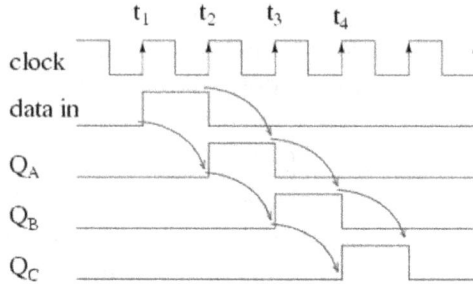

At clock time t_1 a "data in" of 0 is clocked from **D** to **Q** of all three stages. In particular, **D** of stage A sees a logic **0**, which is clocked to Q_A where it remains until time t_2.

At clock time t_2 a "data in" of 1 is clocked from **D** to Q_A. At stages, **B** and **C**, a 0, fed from preceding stages is clocked to Q_B and Q_C.

At clock time t_3 a "data in" of 0 is clocked from **D** to Q_A. Q_A goes low and stays low for the remaining clocks due to "data in" being 0. Q_B goes high at t_3 due to a **1** from the previous stage. Q_C is still low after t_3 due to a low from the previous stage.

Q_C finally goes high at clock t_4 due to the high fed to **D** from the previous stage Q_B. All earlier stages have **0**s shifted into them. And, after the next clock pulse at t_5, all logic **1**s will have been shifted out, replaced by **0**s

Serial In - Parallel Out Shift Registers

For this kind of register, data bits are entered serially in the same manner as discussed in the last section. The difference is the way in which the data bits are taken out of the register. Once the data are stored, each bit appears on its respective output line, and all bits are available simultaneously. A construction of a four-bit serial in - parallel out register is shown below.

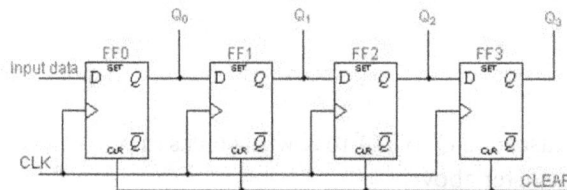

Here, we can see how the four-bit binary number 1001 is shifted to the Q outputs of the register.

CLEAR	Q0	Q1	Q2	Q3
1001	0	0	0	0

The following serial-in/ serial-out shift registers are 4000 series *CMOS* (Complementary Metal Oxide Semiconductor) family parts. As such, they will accept a V_{DD}, positive power supply of 3-Volts to 15-Volts. The V_{SS} pin is grounded. The maximum frequency of the shift clock, which varies with V_{DD}, is a few megahertz. See the full datasheet for details.

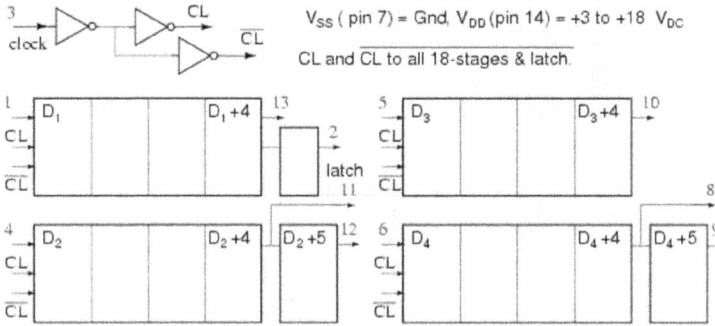

CD4006b Serial-in/ serial-out shift register

The 18-bit CD4006b consists of two stages of 4-bits and two more stages of 5-bits with an output tap at 4-bits. Thus, the 5-bit stages could be used as 4-bit shift registers. To get a full 18-bit shift register the output of one shift register must be cascaded to the input of another and so on until all stages create a single shift register as shown below.

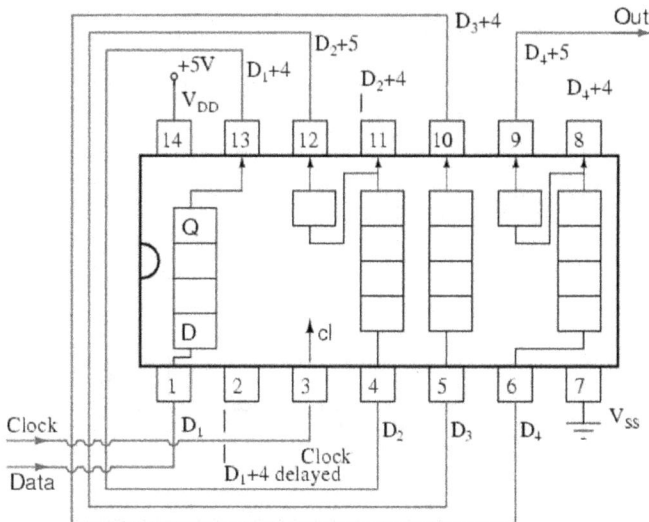

CD4006b 18-bit serial-in/ serial-out shift register

A CD4031 64-bit serial-in/ serial-out shift register is shown below. A number of pins are not connected (nc). Both Q and Q' are available from the 64th stage, actually Q_{64} and Q'_{64}. There is also a Q_{64} "delayed" from a half stage which is delayed by half a clock cycle. A major feature is a data selector which is at the data input to the shift register.

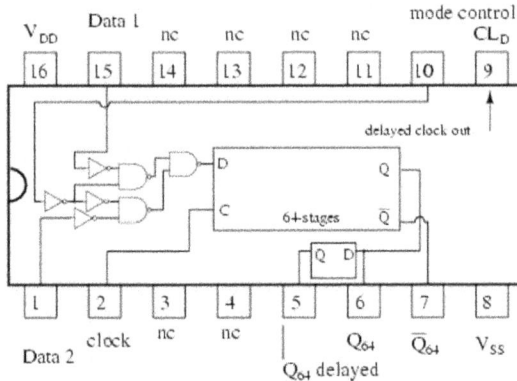

CD4031 64-bit serial-in/ serial-out shift register

The "mode control" selects between two inputs: data 1 and data 2. If "mode control" is high, data will be selected from "data 2" for input to the shift register. In the case of "mode control" being logic low, the "data 1" is selected. Examples of this are shown in the two figures below.

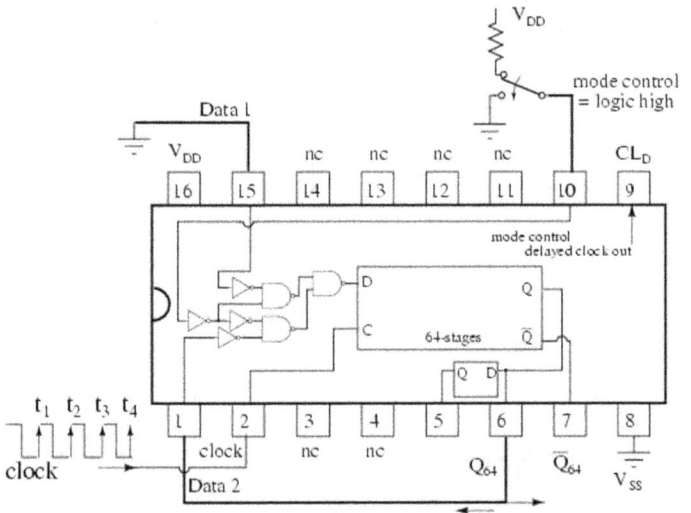

CD4031 64-bit serial-in/ serial-out shift register recirculating data.

The "data 2" above is wired to the Q_{64} output of the shift register. With "mode control" high, the Q_{64} output is routed back to the shifter data input D. Data will *recirculate* from output to input. The data will repeat every 64 clock pulses as shown above. The question that arises is how did this data pattern get into the shift register in the first place?

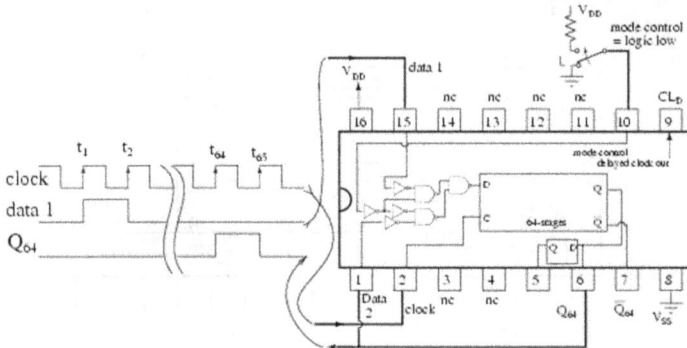

CD4031 64-bit serial-in/ serial-out shift register load new data at Data 1.

With "mode control" low, the CD4031 "data 1" is selected for input to the shifter. The output, Q_{64}, is not recirculated because the lower data selector gate is *disabled*. By disabled we mean that the logic low "mode select" inverted twice to a low at the lower NAND gate prevents it for passing any signal on the lower pin (data 2) to the gate output. Thus, it is disabled.

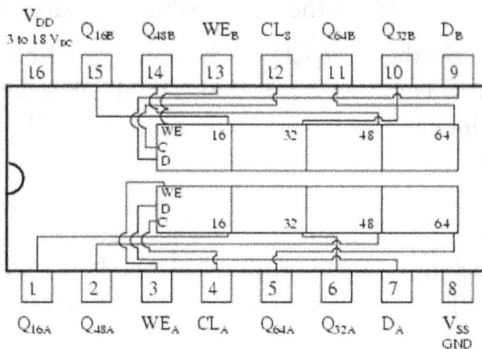

CD4517b dual 64-bit serial-in/ serial-out shift register

A CD4517b dual 64-bit shift register is shown above. Note the taps at the 16th, 32nd, and 48th stages. That means that shift registers of those lengths can be configured from one of the 64-bit shifters. Of course, the 64-bit shifters may be cascaded to yield an 80-bit, 96-bit, 112-bit, or 128-bit shift register. The clock CL_A and CL_B need to be paralleled when cascading the two shifters. WE_B and WE_B are grounded for normal shifting operations. The data inputs to the shift register A and B are D_A and D_B respectively.

Suppose that we require a 16-bit shift register. Can this be configured with the CD4517b? How about a 64-shift register from the same part?

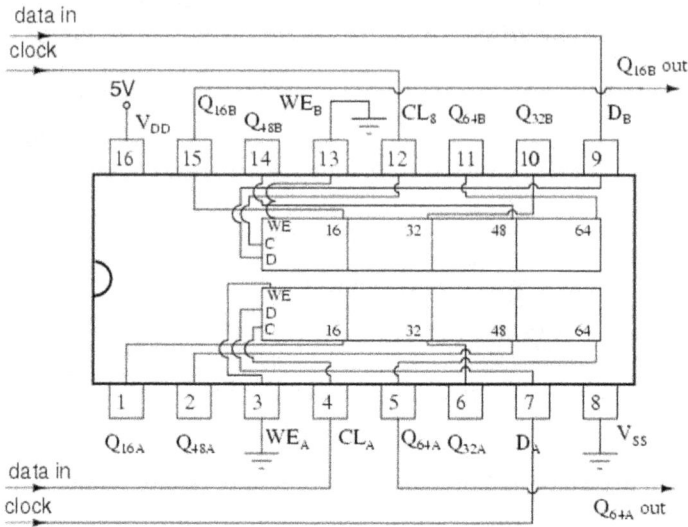

CD4517b dual 64-bit serial-in/ serial-out shift register, wired for 16-shift register, 64-bit shift register

Above we show A CD4517b wired as a 16-bit shift register for section B. The clock for section B is CL_B. The data is clocked in at CL_B. And the data delayed by 16-clocks is picked of Q_{16B}. WE_B, the write enable, is grounded.

Above we also show the same CD4517b wired as a 64-bit shift register for the independent section A. The clock for section A is CL_A. The data enters at CL_A. The data delayed by 64-clock pulses are picked up from Q_{64A}. WE_A, the write enable for section A, is grounded.

Parallel In - Serial Out Shift Registers

A four-bit parallel in - serial out shift register is shown below. The circuit uses D flip-flops and NAND gates for entering data (ie writing) to the register.

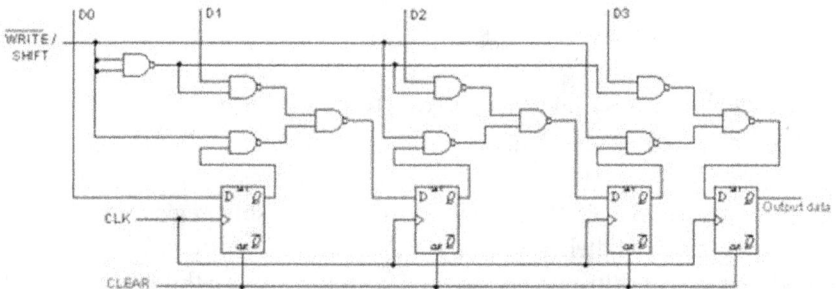

D0, D1, D2, and D3 are the parallel inputs, where D0 is the most significant bit and D3 is the least significant bit. To write data in, the mode control line is taken to LOW and the data is clocked in. The data can be shifted when the mode control line is HIGH as SHIFT is active high. The register performs right shift operation on the application of a clock pulse, as shown in the animation below.

Parallel In - Parallel Out Shift Registers

For parallel in - parallel out shift registers, all data bits appear on the parallel outputs immediately following the simultaneous entry of the data bits. The following circuit is a four-bit parallel in - parallel out shift register constructed by D flip-flops.

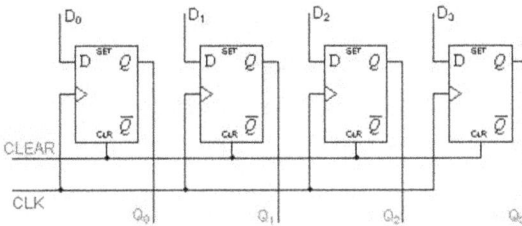

The D's are the parallel inputs and the Q's are the parallel outputs. Once the register is clocked, all the data at the D inputs appear at the corresponding Q outputs simultaneously.

Bidirectional Shift Registers

The registers discussed so far involved only right shift operations. Each right shift operation has the effect of successively dividing the binary number by two. If the operation is reversed (left shift), this has the effect of multiplying the number by two. With suitable gating arrangement, a serial shift register can perform both operations.

A *bidirectional*, or *reversible*, shift register is one in which the data can be shift either left or right. A four-bit bidirectional shift register using D flip-flops is shown below.

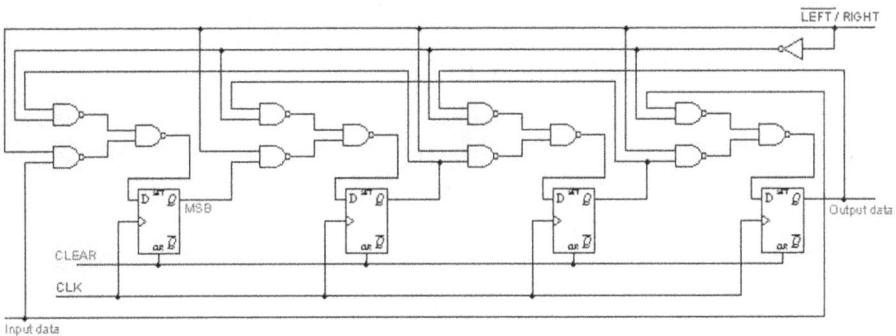

Here a set of NAND gates are configured as OR gates to select data inputs from the right or left adjacent bistables, as selected by the LEFT/RIGHT control line. In this, right shift is done four times and then the left shift is performed four times. Notice the order of the four output bits is not the same as the order of the original four input bits. They are actually reversed!

RIGHT	FF0	FF1	FF2	FF3
11111001	0	0	0	0

Ripple Counter

Definition

An n-stage counter that is formed from n cascaded flip-flops. The clock input to each of the individual flip-flops, with the exception of the first, is taken from the output of the preceding one. The count thus ripples along the counter's length due to the propagation delay associated with each stage of counting.

A ripple counter contains a chain of **flip-flops** with the output of each one feeding the input of the next. A flip-flop output changes state every time the input changes from high to low (on the falling-edge). This simple arrangement works well, but there is a slight delay as the effect of the clock 'ripples' through the chain of flip-flops.

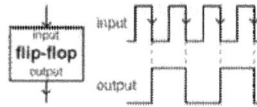

In most circuits, the **ripple delay** is not a problem because it is far too short to be seen on a display. However, a logic system connected to ripple counter outputs will briefly see false counts which may produce 'glitches' in the logic system and may disrupt its operation. For example, a ripple counter changing from 0111 (7) to 1000 (8) will very briefly show 0110, 0100 and 0000 before 1000!

Linking Ripple Counters

The Figure 8.40 shows how to link standard ripple counters. Notice how the highest output QD of each counter drives the **clock (CK)** input of the next counter. This works because ripple counters have clock inputs that are 'active-low' which means that the count advances as the clock input becomes low, on the falling-edge.

Remember that with all ripple counters there will be a slight delay before the later outputs respond to the clock signal, especially with a long counter chain. This is not a problem in simple circuits driving displays, but it may cause glitches in logic systems connected to the counter outputs.

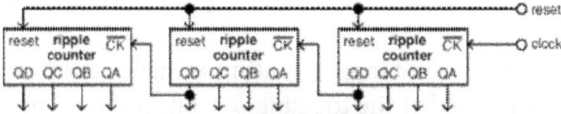

Figure 8.40: Ripple Counters.

Synchronous Counter

Description

A *synchronous counter*, in contrast to an *asynchronous counter*, is one whose output bits change state simultaneously, with no ripple. The only way we can build such a counter circuit from J-K flip-flops is to connect all the clock inputs together so that each and every flip-flop receives the exact same clock pulse at the exact same time (Figure 8.41):

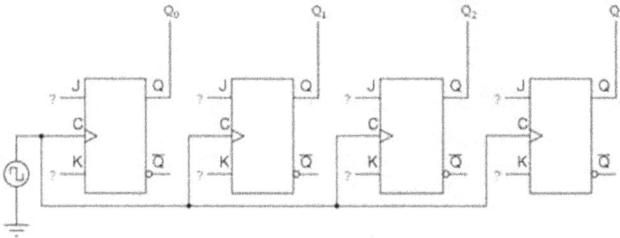

Figure 8.41: A Synchronous Counter.

Operation

Now, the question is, what do we do with the J and K input? We know that we still have to maintain the same divide-by-two frequency pattern in order to count in a binary sequence and that this pattern is best achieved utilizing the "toggle" mode of the flip-flop, so the fact that the J and K inputs must both be (at times) "high" is clear. However, if we simply connect all the J and K inputs to the positive rail of the power supply as we did in the asynchronous circuit, this would clearly not work because all the flip-flops would toggle at the same time: with each and every clock pulse!

This circuit will not function as a counter!

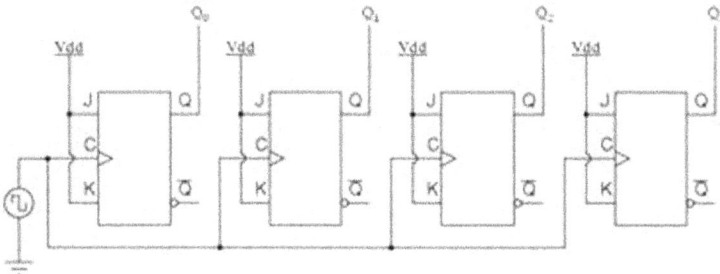

Working

Let's examine the four-bit binary counting sequence again, and see if there are any other patterns that predict the toggling of a bit. Asynchronous counter circuit design is based on the fact that each bit toggle happens at the same time that the preceding bit toggles from a "high" to a "low" (from 1 to 0). Since we cannot clock the toggling of a bit based on the toggling of a previous bit in a synchronous counter circuit (to do so would create a ripple effect) we must find some other pattern in the counting sequence that can be used to trigger a bit toggle. Examining the four-bit binary count sequence, another predictive pattern can be seen. Notice that just before a bit toggles, all preceding bits are "high:"

```
0 0 0 0
0 0 0 1
0 0 1 0
0 0 1 1
0 1 0 0
0 1 0 1
0 1 1 0
0 1 1 1
1 0 0 0
1 0 0 1
1 0 1 0
1 0 1 1
1 1 0 0
1 1 0 1
1 1 1 0
1 1 1 1
```

This pattern is also something we can exploit in designing a counter circuit. If we enable each J-K flip-flop to toggle based on whether or not all preceding flip-flop outputs (Q) are "high," we can obtain the same counting sequence as the asynchronous circuit without the ripple effect since each flip-flop in this circuit will be clocked at exactly the same time:

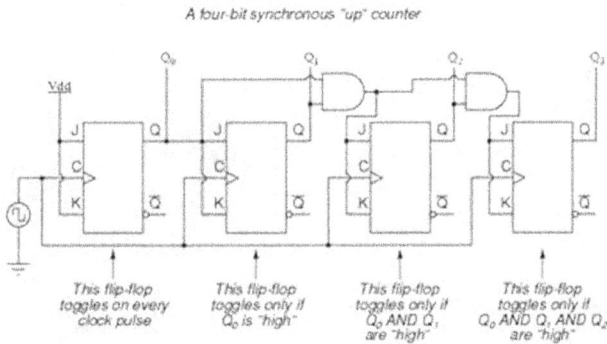

A four-bit synchronous "up" counter

This flip-flop toggles on every clock pulse

This flip-flop toggles only if Q_0 is "high"

This flip-flop toggles only if Q_0 AND Q_1 are "high"

This flip-flop toggles only if Q_0 AND Q_1 AND Q_2 are "high"

The result is a four-bit *synchronous* "up" counter. Each of the higher-order flip-flops is made ready to toggle (both J and K inputs "high") if the Q outputs of all previous flip-flops are "high." Otherwise, the J and K inputs for that flip-flop will both be "low," placing it into the "latch" mode where it will maintain its present output state at the next clock pulse. Since the first (LSB) flip-flop needs to toggle at every clock pulse, its J and K inputs are connected to V_{cc} or V_{dd}, where they will be "high" all the time. The next flip-flop need only "recognize" that the first flip-flop's Q output is high to be made ready to toggle, so no AND gate is needed. However, the remaining flip-flops should be made ready to toggle only when *all* lower-order output bits are "high," thus the need for AND gates.

To make a synchronous "down" counter, we need to build the circuit to recognize the appropriate bit patterns predicting each toggle state while counting down. Not surprisingly, when we examine the four-bit binary count sequence, we see that all preceding bits are "low" prior to a toggle:

```
0 0 0 0
0 0 0 1
0 0 1 0
0 0 1 1
0 1 0 0
0 1 0 1
0 1 1 0
0 1 1 1
1 0 0 0
1 0 0 1
1 0 1 0
1 0 1 1
1 1 0 0
1 1 0 1
1 1 1 0
1 1 1 1
```

Since each J-K flip-flop comes equipped with a Q' output as well as a Q output, we can use the Q' outputs to enable the toggle mode on each succeeding flip-flop, being that each Q' will be "high" every time that the respective Q is "low:"

A four-bit synchronous "down" counter

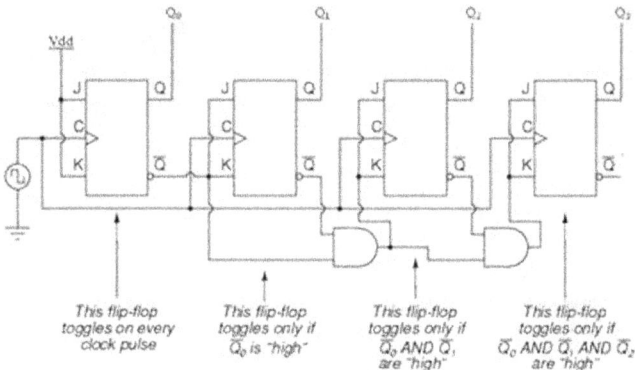

| This flip-flop toggles on every clock pulse | This flip-flop toggles only if $\overline{Q}_0$ is "high" | This flip-flop toggles only if $\overline{Q}_0$ AND $\overline{Q}_1$ are "high" | This flip-flop toggles only if $\overline{Q}_0$ AND $\overline{Q}_1$ AND $\overline{Q}_2$ are "high" |

Taking this idea one step further, we can build a counter circuit with selectable between "up" and "down" count modes by having dual lines of AND gates detecting the appropriate bit conditions for an "up" and a "down" counting sequence, respectively, then use OR gates to combine the AND gate outputs to the J and K inputs of each succeeding flip-flop:

A four-bit synchronous "up/down" counter

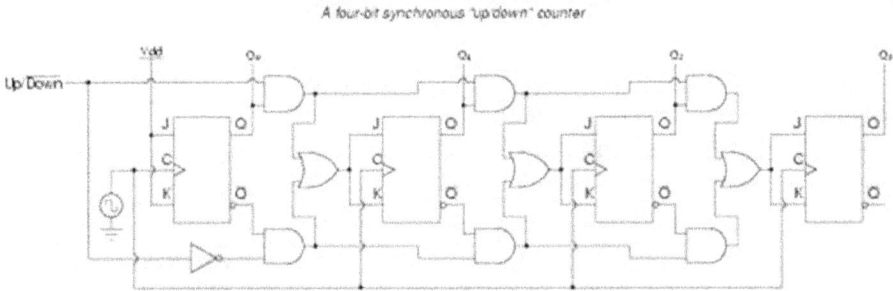

This circuit isn't as complex as it might first appear. The Up/Down control input line simply enables either the upper string or lower string of AND gates to pass the Q/Q' outputs to the succeeding stages of flip-flops. If the Up/Down control line is "high," the top AND gates become enabled, and the circuit functions exactly the same as the first ("up") synchronous counter circuit shown in this section. If the Up/Down control line is made "low," the bottom AND gates become enabled, and the circuit functions identically to the second ("down" counter) circuit shown in this section. To illustrate, here is a diagram showing the circuit in the "up" counting mode (all disabled circuitry shown in grey rather than black):

Counter in "up" counting mode

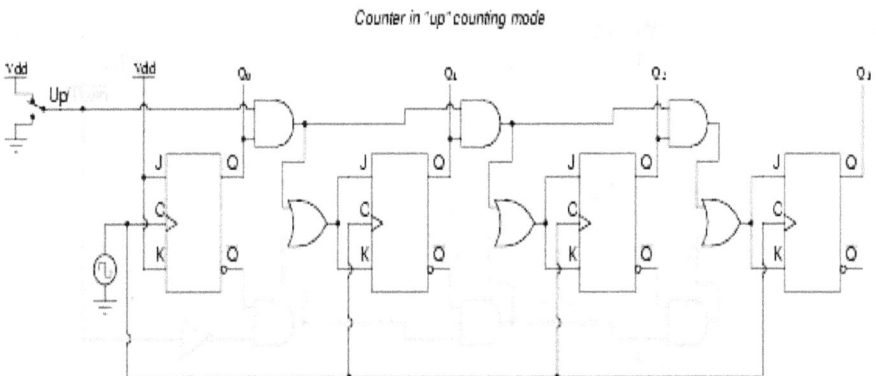

Here, as shown in the "down" counting mode, with the same grey coloring representing disabled circuitry (Figure 8.42):

Counter in "down" counting mode

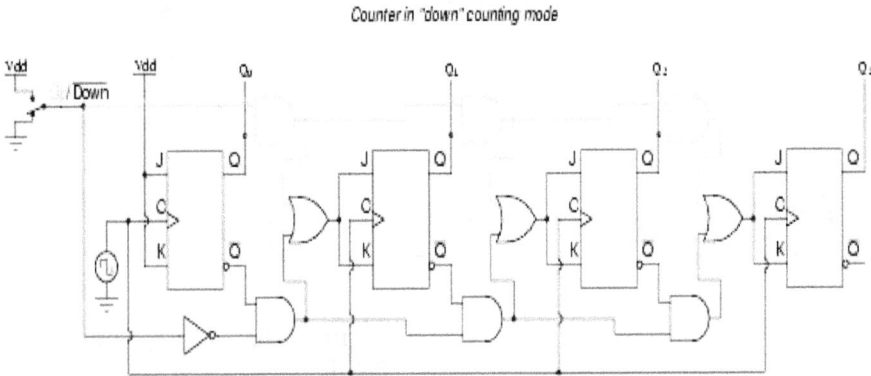

Figure 8.42: Down Counter.

Up/down counter circuits are very useful devices. A common application is in machine motion control, where devices called *rotary shaft encoders* convert mechanical rotation into a series of electrical pulses, these pulses "clocking" a counter circuit to track total motion:

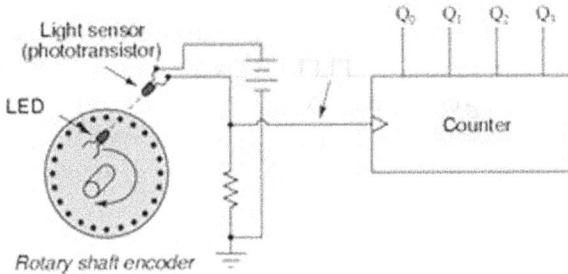

As the machine moves, it turns the encoder shaft, making and breaking the light beam between LED and phototransistor, thereby generating clock pulses to increment the counter circuit. Thus, the counter integrates or accumulates, total motion of the shaft, serving as an electronic indication of how far the machine has moved. If all we care about is tracking total motion, and do not care to account for changes in the *direction* of motion, this arrangement will suffice. However, if we wish the counter to *increment* with one direction of motion and *decrement* with the reverse direction of motion, we must use an up/down counter, and an encoder/decoding circuit having the ability to discriminate between different directions.

If we re-design the encoder to have two sets of LED/phototransistor pairs, those pairs aligned such that their square-wave output signals are 90° out of phase with each other, we have what is known as a *quadrature output* encoder. A phase detection circuit may be made from a D-type flip-flop, to distinguish a clockwise pulse sequence from a counter-clockwise pulse sequence:

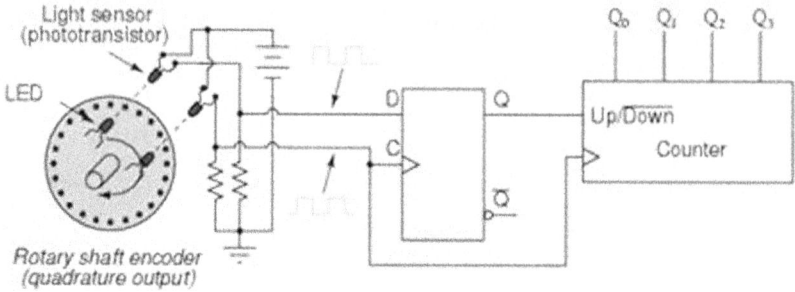

When the encoder rotates clockwise, the "D" input signal square-wave will lead the "C" input square-wave, meaning that the "D" input will already be "high" when the "C" transitions from "low" to "high," thus *setting* the D-type flip-flop (making the Q output "high") with every clock pulse. A "high" Q output places the counter into the "Up" count mode, and any clock pulses received by the clock from the encoder (from either LED) will increment it. Conversely, when the encoder reverses rotation, the "D" input will lag behind the "C" input waveform, meaning that it will be "low" when the "C" waveform transitions from "low" to "high," forcing the D-type flip-flop into the *reset* state (making the Q output "low") with every clock pulse. This "low" signal commands the counter circuit to decrement with every clock pulse from the encoder. This circuit, or something very much like it, is at the heart of every position-measuring circuit based on a pulse encoder sensor. Such applications are very common in robotics, CNC machine tool control, and other applications involving the measurement of reversible, mechanical motion.

Microprocessors

Introduction

A microprocessor—also known as a **CPU** or central processing unit – is complete computation engine that is fabricated on a single chip. The first microprocessor was the Intel 4004, introduced in 1971. The 4004 was not very powerful—all it could do was add and subtract, and it could only do that 4 bits at a time. But it was amazing that everything was on one chip. Prior to the 4004, engineers built computers either from collections of chips or from discrete components (transistors wired one at a time). The 4004 powered one of the first portable electronic calculators.

In the world of personal computers, the terms *microprocessor* and CPU are used interchangeably. At the heart of all personal computers and most workstations sits a microprocessor. Microprocessors also control the logic of almost all digital devices, from clock radios to fuel-injection systems for automobiles.

Characteristics

Three basic characteristics differentiate microprocessors:

Instruction set: The set of instructions that the microprocessor can execute.

Bandwidth: The number of bits processed in a single instruction.

Clock speed: Given in megahertz (MHz), the clock speed determines how many instructions per second the processor can execute.

In case of 32-bit microprocessor running at 50MHz, it is observed that it is more powerful than a 16-bit microprocessor that runs at 25MHz. In addition to bandwidth and clock speed, microprocessors are classified as being either RISC (reduced instruction set computer) or CISC (complex instruction set computer).

If you have ever wondered what the microprocessor in your computer is doing, or if you have ever wondered about the differences between types of microprocessors, then read on. In this article, you will learn how fairly simple digital logic techniques allow a computer to do its job, whether it's playing a game or spell checking a document

Timeline of Microprocessors

Abacus (3000 B.C): was the first computer in the history of the computing machines. 1st Generation Computers (1946-1952): fastest machine of the time. 2nd Generation Computers (1956): transistor was first time used in it. 3rd Generation Computers: integrated (circuit) chips. 4th Generation Computers: Highly sophisticated technology required. Personal Computers (1975): MITS ALTAIR first computer with TV. 1995: Pentium-Pro Microprocessor Windows 95 OS In 1997 Intel announces Multimedia capabilities (PII). 1998 Windows 1998 OS supported processors. This was much user friendly.

Revolution in the Fabrication of Microprocessors

After launching of windows 98(OS) there started a race of higher performance computing machines. Microprocessors started to reform in small and small size but the speed was increasing by and by and this race is still going on and now its momentum is much faster than ever before. There is a detailed history of processors which have been produced since the year 2000 to till now.

There are some big or major companies which are producing microprocessors and running side by side in that race but three companies are most famous for this work 1st is **Intel** and 2nd is **IBM** and 3rd is **AMD**.

Intel

In these companies the most prominent company is **Intel**. It is just because of the reliability and better results of Intel processors. Here is a report on Intel processors. If we start counting from the year 2000 their first processor was based on "Celeron" technology.

Intel Celeron

Introduced in April 1998, the first Celeron branded CPU was based on the Pentium II branded Core. Subsequent Celeron branded CPUs were based on the Pentium III, Pentium 4, Pentium M, and Core 2 Duo branded processors.

The Celeron brand refers to a range of Intel's x86 CPUs for budget/value personal computers. Considered Intel's "economic" processor, the Celeron branded processors have complemented Intel's higher-performance (and more expensive) brands. Intel has given the brand the motto, "delivering great quality at an exceptional value." Celeron processors can run all IA-32 computer programs, but their performance is somewhat lower when compared to similar, but higher priced, Intel CPU brands. For example, the Celeron brand will often have less cache memory, or have advanced features purposely disabled. These missing features have had a variable impact on performance. In some cases, the effect was significant and in other cases the differences were relatively minor. Many of the Celeron designs have achieved a very high "bang to the buck," while at other times, the performance difference has been noticeable.

History of Intel Celeron Processors from the year 2000 up to now.

Year of Production	Proceedings
Year 2000	**January 4** Intel® Celeron® Processor 533 MHz **February 14** Mobile Intel® Celeron® Processor 500 MHz, 450 MHz **June 19** Low Voltage Mobile Intel® Celeron® Processor 500 MHz
Year 2001	**January 3** Intel® Celeron® Processor 800 MHz **October 2** Intel® Celeron® Processor 1.20 GHz
Year 2002	**January 3** Intel® Celeron® Processor 1.30 GHz **November 20** Intel® Celeron Processor 2.20 GHz, 2.10 GHz
Year 2003	**January 14** Mobile Intel® Celeron® Processor 2 GHz Low Voltage Mobile Intel® Celeron® Processor 866 MHz **November 12** Mobile Intel® Celeron® Processor 2.50 GHz Ultra Low Voltage Mobile Intel® Celeron® Processor 800 MHz

Year of Production	Proceedings
Year 2004-07	**January 4, 2004** Intel® Celeron® M Processor 320 and 310 1.3 GHz 1.2 GHz **July 20, 2004** Intel® Celeron® M processor Ultra-Low Voltage 353 900 MHz **March** Intel® Celeron® M Processor 430-450 1.73-2.0GHz **November 23** Intel® Celeron® D Processor 345 3.06 GHz ** No Celeron Processor was released in 2007*
Year 2008	**January 2008** Celeron Core 2 Duo (Allendale)

Pentium

The **Pentium** brand refers to Intel's single-core x86 microprocessor based on the **P5** fifth-generation micro architecture considered here as such only. The name 'Pentium' was derived from the Greek *penta*, meaning 'five', and the Latin ending -*ium*.

Introduced on March 22, 1993 the Pentium succeeded the Intel 486, which number "4" signified the fourth-generation *micro architecture*. In 1996, the original *Pentium* was succeeded by the **Pentium MMX** branded CPUs still based on the P5 fifth-generation microarchitecture.

Starting in 1995, Intel (inconsistently) used the "Pentium" registered trademark in the names of families of post-fifth-generations of x86 processors branded as the Pentium Pro, Pentium II, Pentium III, Pentium 4 and Pentium D (see Pentium (brand)). Although they shared the x86 instruction set with the original Pentium (and its predecessors), their micro architectures were radically different from the P5 micro architecture of CPUs branded just as the "Pentium" and "Pentium MMX".

Here is a brief History of Intel Pentium Processors

Year of Production	Proceedings
Year 2000	**March 20** Intel® Pentium® III Processor 866 MHz, 850 MHz **March 8** Intel® Pentium® III Processor 1 GHz **November 20** Intel® Pentium® 4 Processor 1.50 GHz, 1.40 GHz

Year of Production	Proceedings
Year 2001	**April 23** Pentium® 4 Processor1.7 **July 2** Pentium® 4 Processor 1.80 GHz, 1.60 GHz **August 27** Intel® Pentium® 4 Processor 2 GHz, 1.90 GHz0 GHz
Year 2002	**January 7** Intel® Pentium® 4 Processor 2.20 GHz, 2 GHz **January 8** **Intel® Pentium® III Processor for servers** 1.40 GHz **April 2, 2002** Intel® Pentium® 4 Processor2.40 GHz, 2.20 GHz **January 21** Ultra Low Voltage Mobile Pentium® III Processor-**M** 750 MHz Low Voltage Mobile Pentium® III Processor-**M** 866 MHz, 850 MHz **November 14, 2002** Intel® Pentium® 4 Processor 3.06 GHz with **Hyper-Threading Technology**
Year 2003	Mobile Intel® Pentium® 4 Processor-M 2.40 GHz (**400 MHz PSB**) **May 21,** **Intel® Pentium® 4 Processor with Hyper-Threading Technology** 2.80C GHz, 2.60C GHz, 2.40C GHz **November 3,** **Intel® Pentium® 4 Processor Extreme Edition** 3.20 GHz
Year 2004	**February 2, 2004** Intel® Pentium® 4 Processor (90nm) 3.40 GHz, 3.20 GHz, 3.0 GHz, 2.80 GHz Intel® Pentium® 4 Processor Extreme Edition (0.13 **micron**) 3.40 GHz **April 7, 2004** Ultra Low Voltage Intel® Pentium® M Processor 1.10 GHz, 1.30Ghz **November 15, 2004** Intel® Pentium® 4 Processor Extreme Edition supporting HT Technology 3.46 GHz

Year of Production	Proceedings
Year 2005-06	Intel® Pentium® 4 Processor **Extreme Edition** supporting HT Technology 3.80 GHz (570) **April, 2005** Intel® Pentium® Processor Extreme Edition 840 3.20 GHz * *No Pentium Processor designed in 2006.*
2007 & 2008	Intel® Pentium™ Processor Extreme Edition 955 3.46 GHz Intel® Pentium™ Processor Extreme Edition 965 3.73 GHz * *No Pentium processor designed yet in 2008*

Xeon

The Xeon brand refers to many families of Intel's x86 multiprocessing CPUs – for dual-processor (DP) and multi-processor (MP) configuration on a single motherboard targeted at non-consumer markets of server and workstation computers, and also at blade servers and embedded systems. The *Xeon* brand has been maintained over several generations of x86 and x86-64 processors. Older models added the *Xeon* moniker to the end of the name of their corresponding desktop processor, but more recent models used the name *Xeon* on its own. The *Xeon* CPUs generally have more cache than their desktop counterparts in addition to multiprocessing capabilities. Intel's (non-x86) IA-64 processors are called Itanium, not *Xeon*.

Here is a brief history of Intel Xeon Processors

Year of Production	Proceedings
Year 2000 & 2001	**January 12** Intel® Pentium® III Xeon™ Processor 800 MHz **September 25, 2001** Intel® Xeon™ Processor 2 GHz **May 24** Intel® Pentium® III Xeon™ Processor 933 MHz
Year 2002-04	**January 9, 2002** Intel® Xeon™ Processor 2.20 GHz **March 12, 2002** Intel® Xeon™ **Processor MP**1.60GHz **March 10, 2003** Intel® Xeon™ Processor 3 GHz (400 MHz system bus)

Year of Production	Proceedings
	November 18 Intel® Xeon™ Processor 2.80 GHz, **October 6, 2003** Intel® Xeon™ Processor 3.20 GHz **March 2, 2004** Intel® Xeon™ Processor MP 3 GHz **(4 MB L3 cache)**
Year 2005-08	**March, 2005** Intel® Xeon Processor MP 2.666 - 3.666 GHz **October, 2005** Dual Core Intel® Xeon Processor 2.8 GHz **August, 2006** Dual-Core Intel® Xeon™ 7140M 3.33-3.40 GHz

Itanium

Itanium is the brand name for 64-bit Intel microprocessors that implement the **Intel Itanium architecture** (formerly called **IA-64**). Intel has released two processor families using the brand: the original **Itanium** and the **Itanium 2**. Starting November 1, 2007, new members of the second family are again called *Itanium*. The processors are marketed for use in enterprise servers and high-performance computing systems. The architecture originated at Hewlett-Packard (HP) and was later developed by HP and Intel together.

Itanium's architecture differs dramatically from the x86 architectures (and the x86-64 extensions) used in other Intel processors. The architecture is based on explicit instruction-level parallelism, with the compiler making the decisions about which instructions to execute in parallel. This approach allows the processor to execute up to six instructions per clock cycle. By contrast with other superscalar architectures, *Itanium* does not have the elaborate hardware to keep track of instruction dependencies during parallel execution - the compiler must keep track of these at build time instead.

After a protracted development process, the first Itanium was released in 2001, and more powerful Itanium processors have been released periodically. HP produces most Itanium-based systems, but several other manufacturers have also developed systems based on Itanium. As of 2007, Itanium is the fourth-most deployed microprocessor architecture for enterprise-class systems.

Itanium has now become a leading microprocessor. Itanium has been used with Dell™ as well as with HP systems. Intanium 1 is being upgraded to Itanium 2 by the

Inter Corporation. Itanium 2 will be the giant of micro-processing as it can execute billions of instruction in a second causing the computing to turn a revolutionary change.

Dual Core

The Core brand was launched on January 5, 2006 by the release of the 32-bit **Yonah** core CPU - Intel's first dual-core mobile (low-power) processor. Its dual-core closely resembled two interconnected Pentium M branded CPUs packaged as a single die (piece) silicon chip (IC). Hence, the 32-bit micro architecture of Core branded CPUs - contrary to its name - had more in common with Pentium M branded CPUs than with the following 64-bit Core micro architecture of Core 2 branded CPUs. Despite a major rebreeding effort by Intel starting January 2006, some computers with the Yonah core continued to be marked as Pentium M.

In 2007, Intel began branding the Yonah core CPUs as Pentium Dual-Core intended for lower-end mobile only computers, unlike the 64-bit Core micro architecture CPUs branded as Intel Core 2 Duo (for higher-end computers) and also as Pentium Dual-Core (for lower-end desktops only). In short, the Core brand refers to processors with the "mobile" derivative of 32-bit Intel P6 micro architecture (preceding the Core micro architecture), whereas the Intel Core 2 Duo brand refers to CPUs with the 64-bit Core micro architecture.

AMD

Advanced Micro Devices (NYSE: AMD) is a leading global provider of innovative processing solutions in the computing, graphics and consumer electronics markets. AMD is dedicated to driving open innovation, choice and industry growth by delivering superior customer-centric solutions that empower consumers and businesses worldwide.

Phenom

Phenom is AMD desktop processor line based on the K10 (not "K10h") micro-architecture, or Family 10h Processors, as AMD calls them. Triple-core versions will be the Phenom 8000 series, quad cores in the Phenom 9000 series, and high-end enthusiast versions in the Phenom FX series. AMD considers the quad core Phenoms to be the first "true" quad core design, as these processors are a monolithic multi-core design, unlike Intel's Core 2 Quad series which are a multi-chip module (MCM) design. The processors will be on the Socket AM2+ platform, with the exception of the high-end model which will only be available for Socket F+. The dual core K10 processors will still be named Athlon X2. Throughout the end of 2007 to 2008, AMD is expected to launch several models of the Phenom processor;

- ☆ Change of model nomenclatures
- ☆ Issues

☆ Future models

☆ See also

☆ External links

☆ References

Change of Model Nomenclatures

The model numbers of the new line of processors were changed from the PR system used in its predecessors, the Athlon 64 X2 family. The new model numbering scheme, for later released Athlon X2 processors, is a four digit model number with different family indicator as the first number, while some Athlon X2 processors used BE as prefix (example as Athlon X2 BE-2400) and some Sempron processors use the LE prefix (example Sempron LE-1200), as follows:

Processor series	Indicator
Phenom quad-core (Agena)	9
Phenom triple-core (Toliman)	8
Athlon dual-core (Kuma)	6
Athlon single-core (Lima)	1
Sempron single-core (Sparta)	

AMD Athlon™

Award-winning processors with exceptional performance to meet your digital demands.

The Athlon 64 is an eighth-generation, AMD64 architecture microprocessor produced by AMD, released on September 23, 2003. It is the third processor to bear the name Athlon, and the immediate successor to the Athlon XP. The second processor (after the **Opteron**) to implement AMD64 architecture and the first 64-bit processor targeted at the average consumer, it is AMD's primary consumer microprocessor, and competes primarily with Intel's Pentium 4, especially the "Prescott" and "Cedar Mill" core revisions. It is AMD's first K8, eighth-generation processor core for desktop and mobile computers. Despite being natively 64-bit, the AMD64 architecture is backward-compatible with 32-bit x86 instructions. Athlon 64s have been produced for Socket 754, Socket 939, Socket 940, and Socket AM2.

Sempron

Sempron has been the marketing name used by AMD for several different entry- level desktop CPUs, using several different technologies and CPU socket formats. The Sempron replaced the AMD Duron processor and competes against Intel's Celeron D processor. AMD coined the name from the Latin semper, which means "always, everyday", to denote that the Sempron was the right processor for everyday computing.

Super Computers

Red Storm™ to be assembled in New Mexico as world's fastest Super computer very soon. Sandia™ supercomputer to be world's fastest, yet smaller and less expensive than any competitor ALBUQUERQUE, N.M. — Red Storm will be faster, yet smaller and less expensive, than previous supercomputers, say researchers at the National Nuclear Security Administration's Sandia National Laboratories, where the machine will be assembled (Figure 9.1).

Figure 9.1: Super Computer.

The first quarter of the $90 million, 41.5 teraflops (trillion operations/second) machine should be installed at Sandia by the end of September and fully up and be running by January, says Bill Camp (Sandia's Director of Computation, Computers, Information and Mathematics), who heads the effort to design and assemble the innovative machine. Red Storm, an air-cooled supercomputer, is being developed by Sandia and Cray Inc. using mostly off-the-shelf parts.

Design innovations permit the machine, from concept to assembly, to be completed with unusual rapidity. While manufacturers typically require four to seven years from concept to first product on a new supercomputer, Cray says Red Storm will begin testing at Sandia less than 30 months after conceptual work began.

The main purpose of the machine is work for the U.S. nuclear stockpile: designing new components; virtually testing components under hostile, abnormal, and normal conditions; and helping in weapons engineering and weapons physics. The machine is expected to run ten times as fast as Sandia's ASCI Red computer system on Sandia's important application codes. (ASCI Red held first place on the top-500 list of the world's supercomputers for three-and-one-half consecutive years.)

But the machine, because of its uniquely inexpensive design, may become the center of Cray's future supercomputer line, says Camp. "From Cray's point of view, the approach we're pioneering here is so powerful they may want their next supercomputers to follow suit."

The machine has unique characteristics: it is scalable from a single cabinet (96 processors) to approximately 300 cabinets (30,000 processors). In addition, the system was designed with a unique capability to monitor and manage itself. Much of the cost incurred for the machine is non-recurring engineering design costs.

The machine has 96 processors in each computer cabinet, with four processors to a board. Each processor can have up to eight gigabytes of memory sitting next to it. Four Cray SeaStars — powerful networking chips — sit on a daughter board atop each processor board. All SeaStars talk to each other "like a Rubik cube with lots of squares on each face," says Camp. "Cray SeaStars are about a factor of five faster than any current competing capability."

Messages encoded in MPI (the Message Passage Interface standard) move from processor to processor at a sustained speed of 4.5 gigabytes per second bidirectionally. The amount of time to get the first information bit from one processor to another is less than 5 microseconds across the system. The machine is arranged in four rows of cabinets. There are a total of 11,648 Opteron processors and a similar number of SeaStars.

The SeaStar chip includes an 800 MHz DDR Hypertransport interface to its Opteron processor, a PowerPC core for handling message-passing chores, and a seven-port router (six external ports). SeaStars are linked together to make up the system's 3-D (X-Y-Z axis) mesh interconnect. IBM is fabricating the SeaStar chips using 0.13-micron CMOS technology. Visualization will occur inside the computer itself — a capability unique to Red Storm among supercomputers.

Sandia is a multiprogram laboratory operated by Sandia Corporation, a Lockheed Martin company, for the U.S. Department of Energy's National Nuclear Security Administration. Sandia has major R&D responsibilities in national security, energy and environmental technologies, and economic competitiveness.

Blue Jean by the IBM had been a fastest supercomputer for more than 20 years and is being still used by the IBM, although they are searching and manufacturing a new supercomputer for geological and metrological analysis of Earth.

Intel 8086/8088 Microprocessor

Microprocessor is a processor (CPU) that is contained on a single silicon chip.

1. A Historical Background of Microprocessors (Intel)
   - ☆ Intel 4004: the world's first microprocessor developed by Intel Corporation in 1971.
     - ✓ 4-bit cpu
     - ✓ 4096 4-bit memory locations
     - ✓ 45 instructions
     - ✓ Speed: 50k 1ps (instructions per second)

☆ Other Intel Microprocessors

1. 8008 (later in 1971): and extended 8-bit version of 4004, memory size: 16k bytes, 48 instructions, 50k 1ps.
2. 8080 (1973): 8-bit cpu, 64k byte memory, 500k 1ps.
3. 8088/8086 (1977,1978): 16bit cpu, 1M byte memory, 2.5M 1ps.
4. 80186: a similar version of 8086, it has more instructions.
5. 80286 (1983): an extended 16M-byte memory of 8086, 8MHz cpu, 4M 1ps.
6. 80386 (1986): fully 32-bit cpu, up to 4G byte memory, hardware memory management and memory assignment. Pipelined instruction execution, 33M 1ps.
7. 80486 (DX, SX) (1989): an improvement of 80386, 50M 1ps, with built-in math processor (for floating-point and extended-precision number operations)
8. Pentium (1993): P5 or 80586, 60-133M2, 16k cache, 4G memory, 2 integer units.
9. Pentium with MMX
10. Pentium Pro (1995): P6, 150-166 M2, 16k cache, 156k-second level cache, three integer units.
11. Pentium 4
12. ...

Fundamentals

Introduction

The 8086/8088 refers to microprocessors developed by Intel Corporation, which was used by IBM personal computers (XT, 1981). 8088 and 8086 are functionally identical (8-bit cpu) except that 8088 uses 8 dada lines in its data bus while 8086 uses 16 data lines in its data bus.

Fundamental Components

The 8086/8088 microprocessor consists of four functional units.

1. Execution unit (EU): decodes and executes machine instructions.
2. Arithmetic and logic unit (ALU): performs math and logical operations on command by the EU.
3. Internal storage (sometimes called registers): is used for internal data storage.
4. Bus interface unit (BIU): handles all communications with the I/O via the system bus and maintains instruction queue.

Execution Unit (EU)

The Execution Unit controls the activity within the processor. It is responsible for retrieving binary machine-language instructions from the **Instruction Queue** maintained by the Bus Interface Unit deciphering them, and seeing to it that the correct steps are performed within the processor to carry out each instruction. If an instruction requires the use of data that is stored internally, in a register, then the Execution Unit retrieves the data from the correct register; if the instruction requires external data, from memory perhaps, then the EU requests the data from the Bus Interface Unit.

Whenever an instruction calls for an arithmetic or logical function, the EU passes the data to Arithmetic and Logic Unit together with a command telling the ALU what to do with the data. The EU then accepts the resulting data from the ALU and sees to it that it is stored in correct location (register or memory), as designated by the instruction.

Arithmetic and Logic Unit (ALU)

The ALU contains circuitry that is capable to perform arithmetic ($+$, $-$, $*$, $/$) operations and logic operations (AND, OR, NOT, XOR) operations. As the ALU completes a requested operation, it also controls individual bits of Flags register (sometimes called the program status register). It sets (1) or clear (0) the correct bits to reflect specific characteristics of the result of the operation, such as whether the result was zero or non-zero, whether it was positive or negative, or if the result is too big to be stored in a byte or word. The EU checks the status of those Flag bits whenever executing conditional jump instructions.

Bus Interface Unit (BIU)

BIU is a circuitry to response memory access and communicates with I/O devices. In most microprocessors, these functions are performed by EU, and sometimes the EU has to be idle due to the different speed and serial execution. In the design of the 8086/8088, Intel introduces BIU in order to avoid the idle time and hold an instruction queue such that instructions can be executed in a pipeline fashion.

System Bus and other Support Hardware Devices in IBM PC

In order to build a complete computer, there are a number of other hardware devices (chips) in the IBM PC. Some of them are:

Memory

Up to one megabyte of system memory, including that which is on the system board, the video adapter card, and any add-on memory boards.

Timer chip

Generates an external interrupt every fifty-five milliseconds to allow the PC to keep track of real time (system data and time).

8259 Interrupt Controller chip

Processes hardware (external) interrupts

6845 CRT Controller

Located on the video adapter card. Controls the video signals to the monitor.

NEC D765 or Intel 8272 Floppy Disk Controller

Located on the disk adapter card. Acts as an interface between the processor and the disk drive.

Inter 8237 Direct Memory Access (DMA) Controller

Located on the system board. Used by the disk controller to transfer data between disk and memory.

8250 Asynchronous Communications Element or Universal Asynchronous Receiver/Transmitter (UART)

Located on each serial communications adapter card (COM1, COM2, *etc.*).

System Bus: is a series of conductive traces of signal lines on the system board, which is used to communicate between 8086/8088 (cpu) and all other devices.

The system bus is made up of three functional parts:

1. Data bus
2. Address bus
3. Control bus

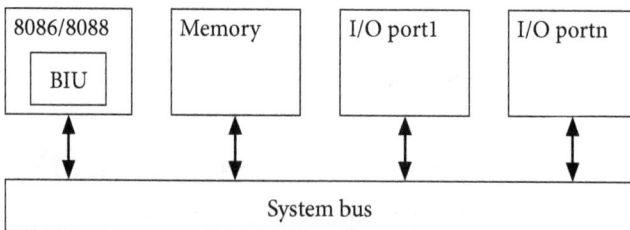

To read from memory, for example, the Bus Interface Unit puts the correct memory address onto the Address Bus and puts the command to read from memory onto the Control Bus. All devices connected to the bus see this address and command simultaneously, but only the memory-control circuitry respond to it. The memory-control circuitry is then responsible for decoding the address, retrieving the data from the appropriate memory chips, and placing the data onto the Data Bus for retrieval by the BIU.

To write to memory, the BIU puts the memory address onto the Address Bus, the byte of data onto the Data Bus, and a command to write data into the memory of the control bus. The memory-control circuitry decodes the command and the address, retrieves the data from the Data Bus, and stores it into the correct memory chips. All other circuitry simply ignores the command.

Communication with the many special-purpose microprocessors attached to the System Bus is accomplished through I/O ports. I/O ports are used for the transfer of data between the 8086/8088 processor and the other support hardware within the system. The In and OUT instructions tell the processor to input or output data through the I/O ports.

When executing an IN instruction, the BIU puts the I/O port address onto the Address Bus and puts the command to input data onto the Control Bus. The circuitry on some peripheral device attached to the bus recognizes the read command and decodes the I/O port as the address of some register within the peripheral device. It then retrieves the data from the register and places it onto the Data Bus from which it is retrieved by the BIU and fed to the Execution Unit within the processor.

To execute an OUT instruction, the BIU puts the command to output data onto the Control Bus, and I/O address onto the Address Bus, and the data to be output onto the Data Bus. The circuitry of a peripheral device is then responsible for recognizing the output command and the I/O address and for retrieving the data from the data bus.

Memory and Internal Storage

Storage Elements

- ☆ Bits: The smallest storage element on any computer is the bit, which can have a value of 0 or 1.
- ☆ Nibbles: A nibble is a sequence of 4 bits.

 e.g.

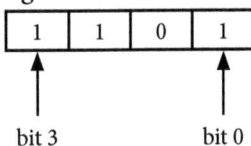

☆ Bytes: A byte is a sequence of 8 bits.

e.g.

1	0	1	0	1	1	0	1

↑ bit 7   ↑ bit 0

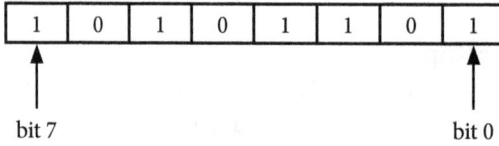

☆ Words: in 8086/8088, a word consists of 16 bits that can be viewed as:

1. A horizontal sequence of 16 bits or.

e.g.

1	0	1	0	1	1	0	1	1	0	1	0	1	1	0	1

↑ bit 15   ↑ bit 0

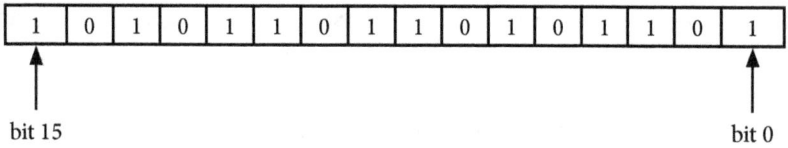

2. A vertical sequence of 2 bytes *e.g.*

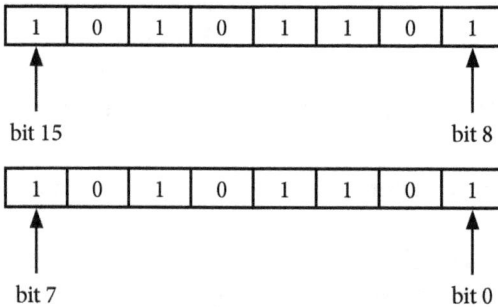

1	0	1	0	1	1	0	1

↑ bit 15   ↑ bit 8

1	0	1	0	1	1	0	1

↑ bit 7   ↑ bit 0

Note: In memory, the least significant bye (bits 0-7) is stored in lower numbered memory location, and the most significant byte (bits 8-15) is stored in the next higher numbered location.

☆ Double words: Two words represents a double word which is viewed as

1. A horizontal sequence of 32 bits or
2. A vertical sequence of 2 words or
3. A vertical sequence of 4 bytes

Note: in memory, its 0-7 bits would come first; bits 8-15 would be in the next higher numbered byte and so on.

☆ Quadword: A quadword is 2 double words, hence 4 words or 8 bytes.

☆ Tenbytes: A tenbyte is a sequence of 10 bytes.

Memory

Computer memory consists of an ordered sequence of storage units (8-bits called bytes), each with its own address.

☆ 8086/8088's memory is bytes-addressable, which means that each byte has its own address, *e.g.* if A is the address of a word (16 bits), then A is actually the address of the first byte (bits 0-7) of the word, A+1 is the address of the second byte (bits 8-15) of the word, and A+2 is the address of the next word.

☆ Memory address space: (MAS): 2^{20}= 1,048,576= 1M(bytes) (address space: 0—2^{20}- 1, *i.e.* 00000-fffff: in hexadecimal)

☆ Map of the memory address space (MAS)

The whole MAS is organized by a $2^{16} * 2^4$ matrix: there are 2^{16} paragraphs, each of which with 16 (2^4) bytes.

Paragraph numbers:

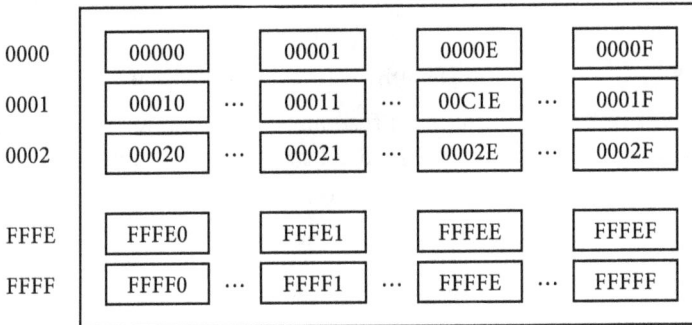

0000	00000	00001	0000E	0000F
0001	00010 ···	00011 ···	00C1E ···	0001F
0002	00020 ···	00021 ···	0002E ···	0002F
FFFE	FFFE0	FFFE1	FFFEE	FFFEF
FFFF	FFFF0 ···	FFFF1 ···	FFFFE ···	FFFFF

Map of the memory address space (MAS) of the 8086/8088.

☆ Hexadecimal number system

The hexadecimal number system uses sixteen characters to represent numbers:0 through 9 and A through F; A is worth ten, B is worth eleven, ..., and F is fifteen. In assembly program, the hexadecimal will be indicated by the letter h.

Example: 7Fh is the number 127($7*16+15*1$) and its equivalent binary number is 01111111b.

The conversion between binary and hexadecimal is simple. To convert from hexadecimal to binary: just represent each hexadecimal digit by four binary digits: Example: 74Dh = 011101001101b

To convert from binary to hexadecimal: break the binary number into groups of four digits, and convert each group into one hexadecimal digit: Example: 101001001110b = A4Fh

☆ Character representation

Each character is represented by a byte or two hexadecimal digits according to ASCII code.

	ASCII code	Letter
For example:	41h	A
	42h	B
	.	.
	.	.
	5Ah	Z
	61h	a
	62h	b
	.	.
	.	.
	7Ah	z
	30h	0
	.	.
	.	.
	39h	9

In assembly, the ASCII representation of character can be defined by enclosing the characters in paired single (') or double (") quotes.

Example: 'A'   'e'   "9"

Note 'A' = 41h, 'a' = 61h and "9" = 39h

☆ Segment-relative addresses.

1. There are 16 bits in a segment register. The content of the segment register represents the paragraph number of the paragraph at which it begins.

2. All memory address is formed by segment-relative format which is computed as follows:

    Absolute address=(segment register)*16 + offset (16-bit) denoted by segment register: where the offset can be content of a register (16-bit) and a 16-bit value.

    e.g.

CS:	1110000000000000
CS*16:	11100000000000000000 (Beginning of code segment)
IP:	0000000000001101 (offset)
--	
CS : IP = 11100000000000001101 (Actual location (absolute address) in memory)	

where CS is the code segment register, IP is the instruction pointer register

Program Stack

A program stack is a region of memory (RAM), which is defined by the assembly program. The program stack is implemented by setting the SS (stack segment) register and by initializing the SP (stack pointer) register to the byte immediately above the top of the sack segment. Data is moved onto the stack in word- sized units by an operation called a push and retrieved from the stack in word-sized units by an operation called pop in the reversed order (last in-first out) (Figure 9.2).

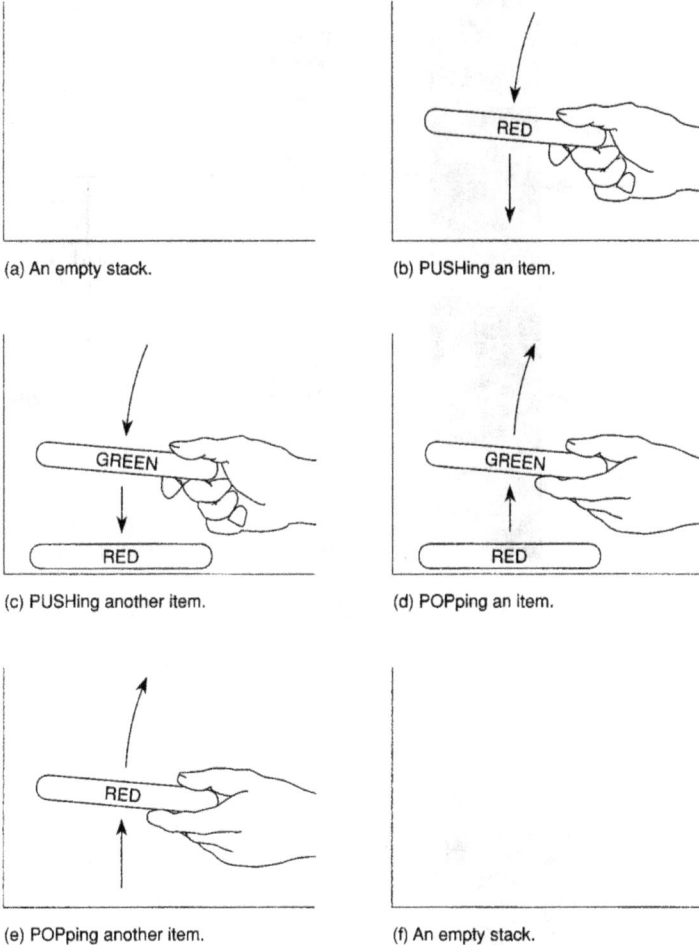

(a) An empty stack.

(b) PUSHing an item.

(c) PUSHing another item.

(d) POPping an item.

(e) POPping another item.

(f) An empty stack.

Figure 9.2: Stack Operations.

A Push operation decrements the SP register twice (−2) and put the word at SS : SP. A Pop operation retrieves the word at SS: SP and increments the SP register twice (+2)

Example: A program stack is to use 100h bytes and SS = 58A1h, A = 1234h and B = 5678h. Figure 9.3 shows the "empty" stack, where the initial value of SP register = 100h and the stack starts at an address of 58A10h and ends at an address of 58B0Fh (100 bytes). Figure 9.4 shows the stack after two Push operations: Push A and Push B, and Figure 9.5 shows the stack after two Pushes and one Pop operations (Pop B).

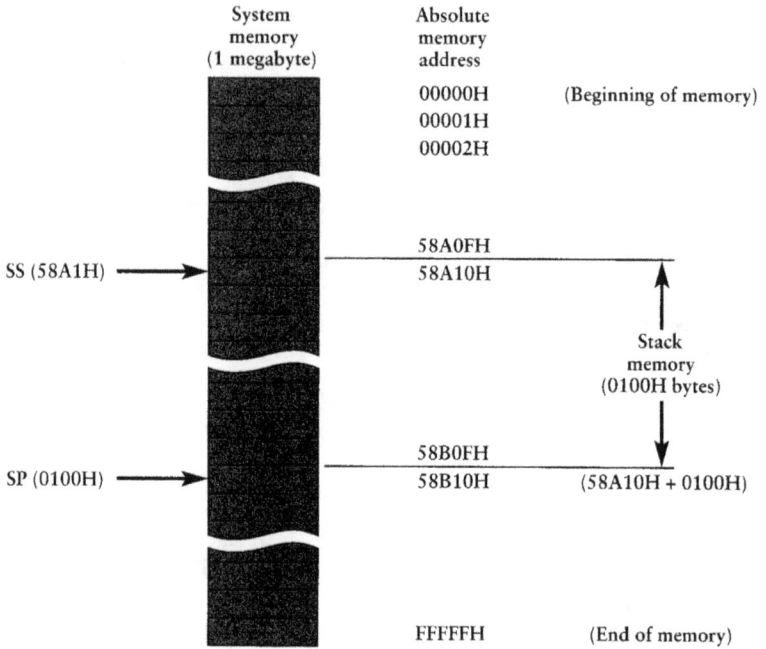

Figure 9.3: Empty Stack.

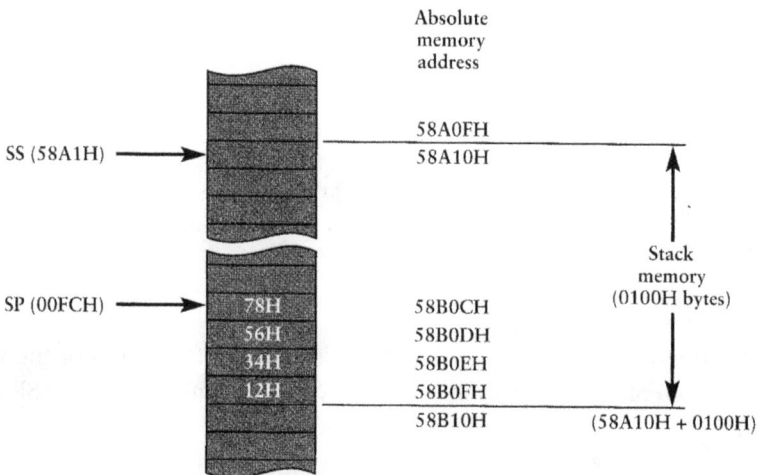

Figure 9.4: Stack after Two Pushes.

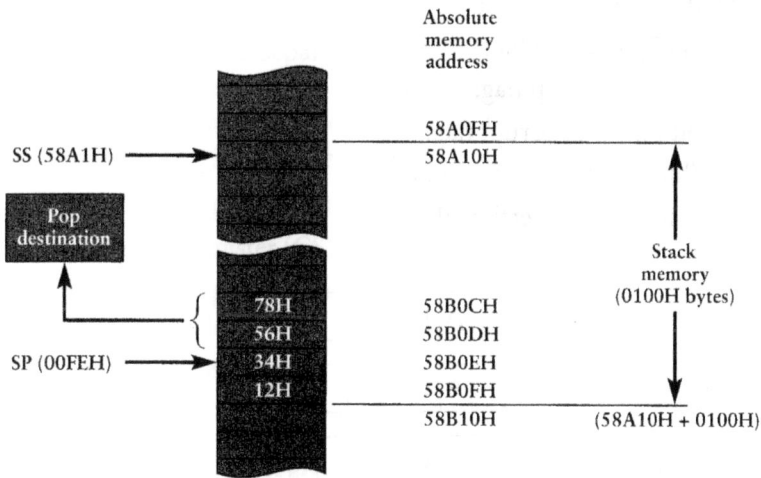

Figure 9.5: Stack after Two Pushes and One Pop.

Internal Storage (Registers)

Registers: there are 14 registers, each 16 bits (bits 0-15) in length. 14 registers are divided into five groups:

1. Four segment registers
2. Four data registers
3. Two pointer and two index registers
4. The instruction pointer register
5. The flag register (status register)

Status Register

In 8086/8088 the status register is called the flag register. The flag register contains information about the most recently executed instruction. There are 16 bits in register. But only bits 0, 2, 4 and 6-11 are used.

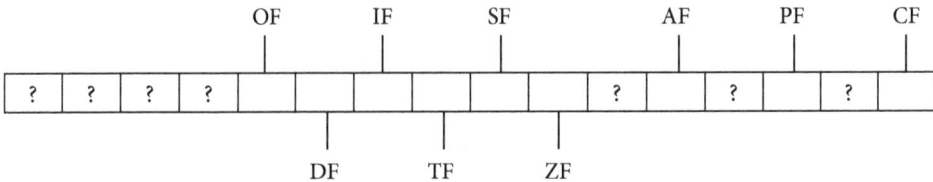

1. CF (bit 0): the carry flag.
2. PF (bit 2): the parity flag.
3. AF (bit 4): the auxiliary flag.

the stack is implemented by a segment of the memory, each item is a 16-bit word (2 bytes). There are two registers associated with stack: SS and SP.

☆ There are two basic stack operations in 8086/8088. a PUSH (operand) it decreases the SP by 1 word (2 bytes), then stores a word on the stack *e.g.* current SP=204. AX=0010. Execution of PUSH AX will result: SP=204-2=202 and store 0010 at memory location at address 202. b POP (operand it removes the top word from the stack, and then add 1 word (2 bytes) to the SP.

☆ The base pointer (BP): BP is also can be used to point to offset within stack segment. Primarily it is used for base indexed addressing modes.

☆ There are two index registers:

1. Source index register (SI)
2. Destination index register (DI) there is no significance to the words source and destination, and SI and DI can be used interchangeably. Index registers are used in the same way subscripts that are used in high-level language. That enable us to access the elements in an array or table.

Instruction Pointer

The instruction pointer (IP) contains the offset address in the code segment of the next instruction to be executed. The absolute address of the next instruction = CS : IP.

Glossary of Terms

Algorithm: A set of mathematical "rules" applied to an input. Generally used to describe a section of computer code which performs a specific function.

Alternating Current (AC): A current whose polarity alternates from positive to negative over time. The rate of such "alternations" is measured in cycles per second - more commonly known as Hertz (Hz).

Amp / Ampere: The basic unit of current flow

Ampere Hour (Amp hour, Ah): A measurement of the capacity of a storage medium (a single cell or a battery). A cell which can supply 1 Amp for 1 hour before it is discharged to a specified minimum level is said to have a capacity of 1 Amp hour.

Amplification: A method for increasing the amplitude (or loudness) of electrical signals.

Amplifier: An electronic device which generates a high power signal based on the information supplied by a lower powered signal. A perfect amplifier would add or subtract nothing from the original except additional power - these have not been invented yet.

Amplitude: The loudness of sound waves and electrical signals. Amplitude is measured in decibels (dB) or volts.

Analogue to Digital Converter (ADC): A device that converts the infinite range of an analogue signal into discrete "steps". Normally, a good audio ADC will use sufficient "steps" to resolve the smallest musical detail. For CD, this is a 16 bit converter, having 65,536 discrete levels covering the most negative signal level to the most positive.

Attenuation: The decrease of a signal's amplitude level over any distance during transmission or through purpose designed attenuators. Attenuation measures signal loss in decibels (dB)

Bandwidth: The measure of a range of frequencies containing an upper and lower limit

Battery: A bank of individual cells connected together to provide the required voltage.

Binary: The basic counting system used in computer logic. Two values are available - 0 and 1.

Binary Code: A coding scheme that communicates information by using a series of "1s" and "Os" that are represented, respectively, by the digital "ON" and "OFF" states.

Bit Stream: The bit rate, or flow of information, between a sender and receiver in digital communication. Also called Digital Bit Stream

Bit: A unit of the binary code that consists of either a single "1" or "O."

Bus: A pathway that connects devices, enabling them to communicate. May be digital or analogue, including power and earth (ground)

Bypass: The practice of using (typically) low value capacitors to conduct high frequency signals either to earth or around a device with limited frequency range

Byte: A unit of the binary code that consists of eight bits. One byte is required to code an alphabetic or numeric character, using an eight-bit character set code

Cable: A type of linear transmission medium.

Capacitor: A pair of parallel "plates" separated by an insulator (the dielectric). Stores an electric charge, and tends to pass higher frequencies more readily than low frequencies. Does not pass direct current, and acts as an insulator. Electrically it is the opposite to an inductor. Basic unit of measurement is the Farad, but is typically measured in micro-farads (uF = 1×10^{-6}F) or nano-farads (nF $- 1 \times 10^{-9}$ F)

Cell: One section of a battery. The common carbon or "alkaline" cells used in battery operated equipment is an example

CMOS: (Complementary Metal Oxide Semiconductor) - one "family" of digital logic devices. Some CMOS devices can operate with power supplies from 3 Volts to 15 Volts - others are limited to the traditional logic 5 Volt power supply.

Coaxial Cable: A metallic cable constructed in such a way that the inner conductor is shielded from EMR (electromagnetic radiation) interference by the outer conductor.

CODEC: COder / DECoder - The component of any digital subsystem which performs analogue to digital and digital to analogue conversions.

Color Code: Used to identify resistors and some capacitors, as well as wires in telephony.

Compression: Component that joins together with a rarefaction to make a sound wave.

Crossover: A filter network which separates frequencies into "bands" which match the capabilities of the loudspeaker drivers within an enclosure.

Crosstalk: A noise impairment when a signal from one pair of wires affects adjacent wires or one channel affects the adjacent channel.

Cutoff Frequency: Normally defined as the frequency where the output from a filter has fallen by 3dB from the maximum level obtainable through the filter.

dB - Decibel - (0.1 Bel): Defined (more or less) as the smallest variation of volume detectable by ear. This is measured on a logarithmic scale, so a change of 3dB from 1 Watt is equivalent to 0.5 Watt or 2 Watts. A change of 10dB from 1 Watt is equivalent to 100mW or 10 Watts. In electronics, 0dBm is a reference value corresponding to 1mW at 600 Ohms - this equates to approximately 775mV. The threshold of sound is 0dB, and typical sounds can reach 140dB or more. Any prolonged sound above 90dB may cause hearing damage.

Digital/Analogue Conversion: A method used to recreate an analogue signal that has been coded into binary data and transmitted as a digital signal.

Digital/Analogue Converter (DAC): A device used to generate a replica of the original analogue signal that has been coded into binary data and transmitted as a digital signal.

Direct Current (DC): A current flow which is steady with time, and flows in one direction only.

Distortion (1): Any modification to a signal which results in the generation of frequencies which were not present in the original.

DSP: Digital Signal Processor - a dedicated computer circuit which performs complex changes or analysis on a digital signal, generally encoded from an analogue source.

Electronic: The use of active electronic components (integrated circuits, transistors, valves *etc.*) which require a power supply to function. Such "active" components will always be used in conjunction with passive components.

Electromagnetic Interference (EMI): An unwanted (possibly interfering) signal emitted by an electronic apparatus. The emission of EMI is heavily regulated in most countries.

Velocity: speed or rapidity. In audio and electronics, we are concerned with the speed of a signal in the air and a conductor. Speed (velocity) of sound in air is approximately 345 metres per second at sea level, but it varies with temperature and humidity. The speed of an electrical signal in a wire is approximately 3 × 10^8 meters per second.

Velocity Factor: A situation that occurs in conductors that are close to another conducting material. For example, a coaxial cable has an inner and outer conductor, with insulation between the two. The velocity factor of such cables varies from 0.7 to 0.9 (*i.e.* the signal travels slower than in free space).

Volt: The basic unit of "electromotive force". One Volt applied to a resistance of one Ohm will force a current of one Ampere to flow (Abbreviation - V).

Watt: The basic unit of power. 1 Volt across 1 Ohm (giving 1 Amp) dissipates 1 Watt (all as heat with a resistive load).

Wavelength: The length of one cycle of an AC signal. Determined by Wavelength = c / f where "c" is velocity and "f" is frequency. The wavelength of a 345Hz audio signal in air is one metre.

Xenon: A gas commonly used in flash tubes, HID (High Intensity Discharge) automotive headlamps, and having an intense white light output with a color temperature close to that of daylight.

Zener Diode: A two-layer device that, above a certain reverse voltage, has a sudden rise in current. If forward biased, the diode is an ordinary rectifier. But, when reversed-biased, the diode exhibits a typical knee, or sharp break, in its current-voltage graph. the voltage remains essentially constant for any further increase of reverse current, up to the allowable dissipation rating.

References

1. Bor, B. Murmain aid B. Digital/v Assisted Pipeline ADCS. Boon, MA: Kiuwer Academic Publishers, 2004.

2. Fbpa C. "Siperior order curvature-corrtion CMOS iiat temperature isor", ASDAM2002, October 2002.

3. Bker-Gom, T.Lakshmi Viswa,ha, aid T.R. Viswaiathai," A Low-Sipply CMOS Sib-Baidg Refere,c, IEEE-TCASII:EXPRESS BRIEFS, Vol. 55 7, July 2008.

4. P. Makovati, *et al.* "Behavior moding of switchi-capacitor sgmad ta modulators," IEEE Trans. Circuits Syst. I, Fundarn. Theoiy App!,50, 3, pp. 352-364, Ma. 2003.

5. A. Becker-Gom "Anog aid Mixi Sgn Thniques for Low Voltage Baidgap PhD distion, 2010.

6. D.A. Neamen, Semiconductor Physics and Devices (IRWIN), Times Mirror High Education Group, Chicago) 1997.

7. E.S. Yang, Microelectronic Devices, McGraw Hill, Singapore, 1988.

8. B.G. Streetman, Solid State Electronic Devices, Prentice Hall of India, New Delhi, 1995.

9. J.Millman and A. Grabel, Microelectronics, McGraw Hill, International, 1987.

10. A.S.Sedra and K.C. Smith, Microelectronic Circuits, Saunder's College Publishing, 1991.

11. R.T. Howe and C.G. Sodini, Microelectronics: An integrated Approach, Prentice Hall International, 1997.

12. J.F.Wakerly: Digital Design, Principles and Practices, 4th Edition, Pearson Education, 2005.

13. Charles H Roth: Digital Systems Design using VHDL, Thomson Learning, 1998.

14. H.Taub and D. Schilling, Digital Integrated Electronics, McGraw Hill, 1977.

15. D.A. Hodges and H.G. Jackson, Analysis and Design of Digital Integrated Circuits, International Student Edition, McGraw Hill, 1983.

16. F.J. Hill and G.L. Peterson, Switching Theory and Logic Design, John Wiley, 1981.

17. Z.Kohavi, Switching and Finite Automata Theory, McGraw Hill, 1970.

18. J.V.Wait, L.P. Huelsman and GA Korn, Introduction to Operational Amplifier theory and applications, 2nd edition, McGraw Hill, New York, 1992.

19. J. Millman and A. Grabel, Microelectronics, 2nd edition, McGraw Hill, 1988.

20. P.Horowitz and W. Hill, The Art of Electronics, 2nd edition, Cambridge University Press, 1989.

21. A.S.Sedra and K.C. Smith, Microelectronic Circuits, Saunder's College Publishing, Edition IV.

22. Paul R.Gray & Robert G.Meyer, Analysis and Design of Analog Integrated Circuits, Wiley, 3 rd Edition.

23. R.S.Gaonkar, Microprocessor Architecture: Programming and Applications with the 8085/8080A, Penram International Publishing, 1996.

24. D A Patterson and J H Hennessy, "Computer Organization and Design The hardware and software interface. Morgan Kaufman Publishers.

25. Douglas Hall, Microprocessors Interfacing, Tata McGraw Hill, 1991.

26. Kenneth J. Ayala, The 8051 Microcontroller, Penram International Publishing, 1996.

27. J.V.Wait, L.P. Huelsman and GA Korn, Introduction to Operational Amplifier theory and applications, 2nd edition, McGraw Hill, New York, 1992.

28. J.Millman and A. Grabel, Microelectronics, 2nd edition, McGraw Hill, 1988.

29. P.Horowitz and W. Hill, The Art of Electronics, 2nd edition, Cambridge University Press, 1989.

30. A.S. Sedra and K.C. Smith, Microelectronic Circuits, Saunder's College Publishing, 1991.

Index

31. D.A. Neamen, Semiconductor Physics and Devices (IRWIN), Times Mirror High Education Group, Chicago) 1997.

32. E.S. Yang, Microelectronic Devices, McGraw Hill, Singapore, 1988.

33. B.G. Streetman, Solid State Electronic Devices, Prentice Hall of India, New Delhi, 1995.

34. J.Millman and A. Grabel, Microelectronics, McGraw Hill, International, 1987.

35. A.S. Sedra and K.C. Smith, Microelectronic Circuits, Saunder's College Publishing, 1991.

36. R.T. Howe and C.G. Sodini, Microelectronics: An integrated Approach, Prentice Hall International, 1997.

www.ingramcontent.com/pod-product-compliance
Lightning Source LLC
Chambersburg PA
CBHW031949180326

41458CB00006B/1665